COST CM0603-MELUSYN Joint Meeting 2010

Damages Induced in Biomolecules by Low and High Energy Radiations

Journal of Physics: Conference Series Volume 261

Paris, France
9 – 12 March 2010

Editors:

Chantal Houee-Levin
Yann A. Gauduel

ISBN: 978-1-61782-297-1
ISSN: 1742-6588

Printed from e-media with permission by:

Curran Associates, Inc.
57 Morehouse Lane
Red Hook, NY 12571

Some format issues inherent in the e-media version may also appear in this print version.

Printed by Curran Associates, Inc. (2011)

For permission requests, please contact the Institute of Physics
at the address below.

Institute of Physics
Dirac House, Temple Back
Bristol BS1 6BE UK

Phone: 44 1 17 929 7481
Fax: 44 1 17 920 0979

techtracking@iop.org

Additional copies of this publication are available from:

Curran Associates, Inc.
57 Morehouse Lane
Red Hook, NY 12571 USA
Phone: 845-758-0400
Fax: 845-758-2634
Email: curran@proceedings.com
Web: www.proceedings.com

TABLE OF CONTENTS

Low Dose Gamma Irradiation Enhances Defined Signaling Components of Intercellular Reactive Oxygen-mediated Apoptosis Induction 1
G Bauer

UVA-induced Damage to DNA and Proteins: Direct Versus Indirect Photochemical Processes 15
P M Girard, S Francesconi, M Pozzebon, D Graindorge, P Rochette, R Drouin, E Sage

Oxidation of Protein Tyrosine Or Methionine Residues: from the Amino Acid to the Peptide 25
J Bergès, P Trouillas, C Houée-Levin

Photosensitization of Plasmid-DNA Loaded with Platinum Nano-particles and Irradiated by Low Energy X-rays 33
E Porcel, K Kobayashi, N Usami, H Remita, C Le Sech, S Lacombe

Bacterial and Archaeal Resistance to Ionizing Radiation 40
F Confalonieri, S Sommer

Synergy Between Low and High Energy Radical Femtochemistry 55
Y A Gauduel

O_2 and Glutathione Effects on Water Radiolysis: a Simulation Study 67
A Colliaux, B Gervais, C Rodriguez-Lafrasse, M Beuve

NEXAFS : A Unique Tool to Follow the Photochemistry of Small Organic Molecules in Condensed Water 78
Ph Parent, C Laffon, F Bournel, J Lasne, S Lacombe

Femtosecond Fluorescence Studies of DNA/RNA Constituents 85
T Gustavsson, A Banyasz, R Improta, D Markovitsi

Radiation Damage to DNA-protein Complexes 98
M Spotheim-Maurizot, M Davídková

Effects of Reactive Oxygen Species on Metabolism Monitored by Longitudinal ^1H Single Voxel MRS Follow-up in Patients with Mitochondrial Disease Or Cerebral Tumors 108
J M Constans, S Collet, J S Guillamo, G Hossu, S Lacombe, Y A Gauduel, C Houée Levin, W Dou, S Ruan, L Barré, F Rioult, J M Derlon, E Lechapt-Zalcman, S Valable, F Chapon, P Courtheoux, V Fong, F Kauffmann

Ion Microbeam Irradiation for Radiobiology and Radical Chemistry: Status and Prospect 125
H Khodja

In Situ UV-visible Spectrum Acquisition of Br_3-. Investigations of Concentrated HBr Aqueous Solutions Under 13-keV X-rays 132
D Saffré, E Atinault, S Pin, J P Renault, J L Hazemann, G Baldacchino

Author Index

PREFACE

The purpose of the four day COST CM0603–MELUSYN Joint Meeting *'Damages induced in Biomolecules by Low and High Energy Radiations'*, held in March 2010 in Paris, France, was to create a timely forum for multi-disciplinary discussions related to the recent developments on biomolecular damage triggered by low or high energy ionising radiations (photons, relativistic particles and accelerated ions).

Created in September 2007, the European network COST CM0603 *'Free radicals in Chemical Biology'* (http://www.chembiorad.inct.pl/index.php) gathers specialists of free radical chemistry, biochemistry and biology. It aims to promote a chemical biology approach for the investigation of free radical pathways of biological interest and the improvement of communication and exchange among neighbouring scientific fields, such as chemistry with several domains of life sciences, specifically addressing the real barriers of specialist language and tools.

The MELUSYN network was born in January 2006. This French think-tank and task force involves about one hundred scientists and clinicians who contribute to the emerging interdisciplinary domain of advanced spatio-temporal radiation biomedicine. One important mission of the MELUSYN network is to explore and identify new medical fields to which the specificity of 3rd generation synchrotron sources could be applied, either by themselves or in synergy with other advanced radiation sources. The network also contributes to the organization of international multidisciplinary conferences.

In the framework of experimental and theoretical approaches, the conference gathered together 50 scientists from both communities and addressed a number of highly relevant aspects of bioradical chemistry and radiation biology, considering classical and more advanced radiation sources (synchrotron, ultra-short particle bunches based on high power laser technologies, microbeams) operating in the broad energy range eV–MeV, classical or semi-quantum simulations of molecular radiation processes. Considering some physical, chemical and biological aspects of radical mechanisms, including their consequences in medicine (radiotherapy for instance), the selected contributions of this Special Issue provide guidance for future interdisciplinary developments. In this way, the multi-scale aspects of radiation based bioradical chemistry and radiation biology should be considered.

We would like to express our thanks to all the authors for their time and genuine efforts, and to the reviewers for their fruitful comments during the preparation of this volume.

Editors

Chantal Houée-Levin
University of Paris Sud, Orsay, France

Yann A Gauduel
Ecole Polytechnique – ENS Techniques Avancées
Palaiseau, France

Journal of Physics: Conference Series **261** (2011) 012001 doi:10.1088/1742-6596/261/1/012001

Low dose gamma irradiation enhances defined signaling components of intercellular reactive oxygen-mediated apoptosis induction

G Bauer
Abteilung Virologie, Institut für Medizinische Mikrobiologie und Hygiene, Universität Freiburg, Freiburg, Germany

E-mail: georg.bauer@uniklinik-freiburg.de

Abstract. Transformed cells are selectively removed by intercellular ROS-mediated induction of apoptosis. Signaling is based on the HOCl and the NO/peroxynitrite pathway (major pathways) and the nitryl chloride and the metal-catalyzed Haber-Weiss pathway (minor pathways). During tumor progression, resistance against intercellular induction of apoptosis is acquired through expression of membrane-associated catalase. Low dose radiation of nontransformed cells has been shown to enhance intercellular induction of apoptosis. The present study was performed to define the signaling components which are modulated by low dose gamma irradiation. Low dose radiation induced the release of peroxidase from nontransformed, transformed and tumor cells. Extracellular superoxide anion generation was strongly enhanced in the case of transformed cells and tumor cells, but not in nontransformed cells. Enhancement of peroxidase release and superoxide anion generation either increased intercellular induction of apoptosis of transformed cells, or caused a partial protection under specific signaling conditions. In tumor cells, low dose radiation enhanced the production of major signaling components, but this had no effect on apoptosis induction, due to the strong resistance mechanism of tumor cells. Our data specify the nature of low dose radiation-induced effects on specific signaling components of intercellular induction of apoptosis at defined stages of multistep carcinogenesis.

1. Introduction

Activation of membrane-associated NADPH oxidase (NOX-1) is functionally connected to oncogenic transformation [1-5]. Extracellular superoxide anions generated by NOX-1 and their dismutation product hydrogen peroxide are crucial for the control of proliferation, maintenance of the transformed state and tumor progression [1-3,5-8]. However, superoxide anions also drive specific elimination of transformed cells through reactive oxygen species (ROS)-mediated intercellular induction of apoptosis [9-15]. [In this paper the term 'ROS' comprises classical reactive oxygen species like superoxide anions, HOCl, hydrogen peroxide or hydroxyl radicals as well as reactive nitrogen species like nitric oxide (NO) or peroxynitrite]. During ROS-mediated intercellular induction of apoptosis, superoxide anion-generating transformed cells represent the 'target cells', whereas nontransformed 'effector cells' contribute to intercellular induction of apoptosis through the release of a novel peroxidase (POD) and nitric oxide. As transformed cells also exhibit effector cell functions (release of POD and/or NO), the interaction of target and effector functions of transformed cells can establish autocrine, ROS-mediated apoptotic signaling [16-18]. The interaction of nontransformed and transformed cells, as well as the autocrine interaction of transformed cells establishes the same four ROS signaling pathways that cause

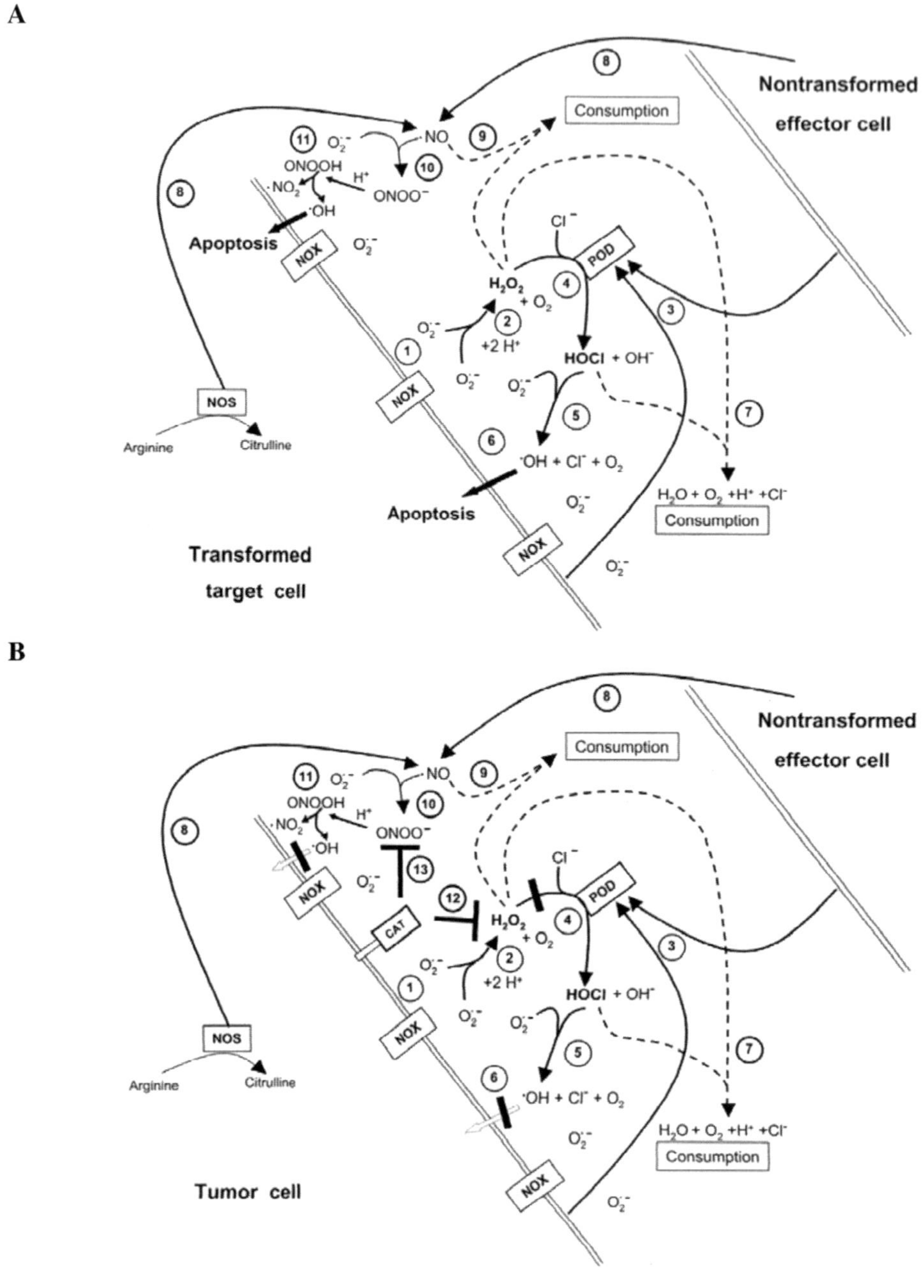

Figure 1: Intercellular ROS-mediated apoptotic signalling. **A** Intercellular ROS signaling between nontransformed and transformed cells and autocrine apoptosis induction in transformed cells. **B**. Catalase protects tumor cells against intercellular and autocrine ROS-mediated apoptosis signaling. Tumor cells show membrane-bound catalase on their surface [16,18]. Catalase, in its classical reaction, decomposes hydrogen peroxide (#12) and thus blocks HOCl signaling (as shown in the figure), as well as the nitryl chloride pathway and the metal-catalyzed Haber-Weiss reaction (not shown in the figure). Catalase also efficiently decomposes peroxynitrite (#13) and thus blocks the NO/peroxynitrite pathway. In addition, compound I of catalase can oxidize NO to NO_2 and thus counteract peroxynitrite formation.

selective apoptosis in transformed target cells (for review see [13-15]). Thereby, the HOCl [11] and the NO/peroxynitrite signaling pathway [4, 11] are the dominant pathways, whereas the nitryl chloride pathway [19] and the metal-catalyzed Haber-Weiss reaction [20] seem to be of minor importance. The details of ROS signaling via the HOCl and the NO/peroxynitrite pathway are shown in figure 1A. Transformed target cells exhibit activated NADPH oxidase (NOX-1) in their membrane and generate extracellular superoxide anions (#1). Superoxide anions dismutate spontaneously to hydrogen peroxide and oxygen (#2). Transformed as well as nontransformed effector cells release duox-coded peroxidase (POD) (#3). This release is triggered by TGF-beta (not shown in the figure). POD utilizes hydrogen peroxide and chloride anions to synthesize HOCl (#4). HOCl then interacts with superoxide anions in the vicinity of the transformed target cell membrane, thereby yielding hydroxyl radicals (#5) that cause lipid peroxidation in the membrane (#6). Lipid peroxidation is followed by ceramide generation, activation of the mitochondrial pathway of apoptosis, activation of caspases 9 and 3, followed by apoptosis (details are not shown in this figure). In the case of excess hydrogen peroxide compared to POD, hydrogen peroxide and HOCl interact in the consumption reaction shown under (#7). This consumption reaction inhibits HOCl signaling. NO synthase (NOS; either iNOS, nNOS or eNOS, depending on the cell system) in target and effector cells utilizes the substrate arginine to synthesize NO (#8). NO readily permeates cellular membranes. It may be consumed by hydrogen peroxide in a complex consumption reaction (#9) [52] that will be described in detail elsewhere. Alternatively, NO interacts with superoxide anions to form peroxynitrite (#10). Peroxynitrite can be protonated to peroxynitrous acid that spontaneously decomposes into NO_2 and apoptosis inducing hydroxyl radicals (#11). For clarity of the graph, the two minor signaling pathways, i.e. the nitryl chloride pathway and the metal-catalyzed Haber-Weiss reaction have not been included. The nitrylchloride pathway utilizes HOCl that interacts with nitrite to form apoptosis-inducing nitryl chloride. The metal-catalyzed Haber-Weiss reaction depends on hydrogen peroxide that reacts with ferrous ions to yield hydroxyl radicals and hydroxyl anions. The resulting ferric ions are then reduced by superoxide anions back to ferrous ions that undergo the next cycle.

Whereas transformed cells are regularly sensitive for intercellular ROS signaling, ex vivo tumor cells exhibit resistance against this process [16-18,21]. Their resistance was shown to be based on interference of catalase located at the outside of the cell membrane with intercellular ROS signaling [16-18] (figure 1B). Expression of membrane-associated catalase was found to be a regular feature of *ex vivo* tumor cells and seems to be required for tumor progression [22,23, Bauer et al., in preparation].

Intercellular effects during oncogenesis may either contribute to tumor formation [24-26] or they may alternatively cause selective elimination of transformed cells through ROS-mediated intercellular induction of apoptosis [9,11,13,27]. Interestingly, both types of intercellular effects seem to be affected by radiation [24,28-30]. Radiation affects carcinogenesis in complex ways and at multiple steps [31]. There is an ongoing discussion on the effects of low dose radiation during carcinogenesis [31,32]. Potential controlling effects of low dose radiation on transformation and tumorigenesis have been well established in the literature [33-44]. Low dose radiation has been shown to induce a decrease in observable transformation frequency [33-36.41-43]. Removal of transformed cells through apoptosis has been discussed as a conceivable mechanism for nontargeted antioncogenic low dose radiation effects [30,45,46]. The report by Portess et al. [30] gives a mechanistic explanation for *selective* elimination of transformed cells by low dose radiation. Their results show that low dose radiation (both of high and low linear energy transfer (LET)) of nontransformed cells caused a marked increase in apoptosis induction in transformed cells [30]). Doses as low as 0.2 mGy alpha particles and 2 mGy gamma ray were sufficient to show distinct effects. Low dose radiation–induced enhancement of intercellular induction of apoptosis was due to the utilization of the same ROS-mediated intercellular pathways as in classical intercellular induction of apoptosis, *i.e.* primarily the HOCl/hydroxyl radical and the NO/peroxynitrite signaling pathway. At this point of the study, the analysis did not allow to define which signaling components were actually enhanced through low dose radiation. The use of neutralizing antibodies to TGF-beta-1 revealed a central role of this cytokine during low dose radiation enhancement of intercellular induction of apoptosis. As only a minority of nontransformed cells had been actually hit by radiation and as TGF-beta shows the ability for autoinduction, Portess et al. suggested a model in which nontransformed cells hit by low dose radiation respond both by release of effector molecules and TGF-beta. TGF-beta then triggers neighbouring cells (that had not been hit by radiation) to propagate TGF-beta signaling and effector molecule release.

The decrease in detectable transformation frequency and removal of transformed cells through low dose radiation effects (frequently termed "protective apoptosis mechanism (PAM)") has become a central element in models for the mathematical simulation of multistage oncogenesis [47-51].
In this paper we present data on the specific effects of low dose gamma irradiation on intercellular induction of apoptosis and autocrine apoptotic selfdestruction and the resultant consequences for the survival of cells at defined stages of multistep oncogenesis.

2. Materials and Methods

Nontransformed murine 208F, transformed 208Fsrc3 and the murine fibrosarcoma cell L929 have been described in reference[16]. The human gastric carcinoma cell line MKN-45 has been described in reference [18]. Media, culture conditions and assays for the quantiation of intercellular and autocrine ROS-mediated apoptosis induction have been described in references [16,18]. These references also describe the sources, concentrations of stock solutions and concentrations of inhibitors used for the characterisation of specific signaling pathways.

Cells (200 000 cells/ml in suspension) were irradiated with a Compagnie Oris (Gif-sur-Yvette, France) IBL 437 Cs^{137} gamma source. The dose rate was between 35 and 40 mGy/sec for the experiments shown in this study. Immediately after irradiation, the cells were centrifuged and resuspended in fresh medium. Cells were then either used directly in the experiments or incubated for an additional hour at 37 °C in a humidified incubator, centrifuged and resuspended in fresh medium.
All other methods are described in the respective figure legends.

3. Results

A novel peroxidase plays a central role for intercellular ROS signaling (Heinzelmann et al., in preparation). Modulation of its release by low dose radiation might explain the recently published effect of low dose radiation on the effector function during intercellular induction of apoptosis [30]. For a direct test of this hypothesis, a new peroxidase test that is based on competition between the catalase/peroxidase mimetic EUK-8 and peroxidase was utilized. This test has been evaluated with myeloperoxidase (MPO) as gold standard and will be published elsewhere. As can be seen in figure 2 A-C, untreated nontransformed fibroblasts, transformed fibroblasts as well as fibrosarcoma cells only released minor concentrations of POD. Nontransformed 208F cells (A), src-oncogene transformed 208Fsrc3 cells (B) and fibrosarcoma cells L929 (C) were irradiated with 40 mGy of gamma radiation or not. Cells were then seeded in Costar 24 well tissue culture clusters (20 000 cells/well) in 0.5 ml complete medium. Some cultures received 20 ng/ml TGF-beta or 10 μM of the matrix metalloprotease inhibitor galardin (Sigma-Aldrich, Deisenhofen, Germany), as indicated in the figures. After 24 hours incubation at 37 °C, supernatants were collected, centrifuged for the removal of cells and stored at -20 °C until they were tested for the presence of POD. For testing the POD content, increasing concentrations of the supernatants were added to 12 500 208Fsrc3 cells in 100 μl complete medium in the presence of 20 ng/ml TGF-beta and 75 μM EUK-8 (Calbiochem). Control assays either were kept free of EUK-8 or received 150 μM of the peroxidase inhibitor aminobenzoyl hydrazide (ABH). All assays were performed in duplicate. Assays were incubated for 7 hours (A, B) or 4.5 hrs (C) at 37 °C and then the percentage of apoptotic cells was determined by the classical apoptosis criteria nuclear condensation, nuclear fragmentation and membrane blebbing, using inverted phase contrast microscopy. A minimum of 200 cell was monitored for each individual assay. Controls without EUK-8 showed less than 7 % apoptotic cells (data not shown). EUK-8 induced between 30 and 50 %

apoptotic cells. This reaction depends on HOCl synthesis by EUK-8 followed by HOCl interaction with 208Fsrc3 cell-derived superoxide anions, resulting in hydroxyl radical mediated apoptosis induction. Natural peroxidase present in the supernatants has been found to interfere with EUK-8-mediated apoptosis induction in a complex way of interaction that will be described elsewhere (Ophoven and Bauer, in preparation). 5- 10 mU/ml myeloperoxidase (MPO) caused 50 % competition. Therefore, competition with EUK-8-mediated apoptosis was indicative for the presence of peroxidase and allowed quantitation. The specificity of the competition reaction was assured by abrogation of competition by the mechanism-based peroxidase inhibitor ABH. This control reaction was performed in all assays. For reasons of simplicity of the graphs, this control is only shown under C.

Treatment of the cells with TGF-beta caused a massive increase in POD release in all three cell systems tested, which resulted in substantial competition. Low dose radiation had the same effect as TGF-beta and proved to be a massive inducer of POD release. In the presence of the matrix metalloprotease inhibitor galardin, no radiation-induced release was observed, indicating that the peroxidase is removed through protease action. The competition by cell-derived peroxidase was abrogated by the mechanism-based peroxidase inhibitor ABH, thus confirming the specificity of the competition reaction.

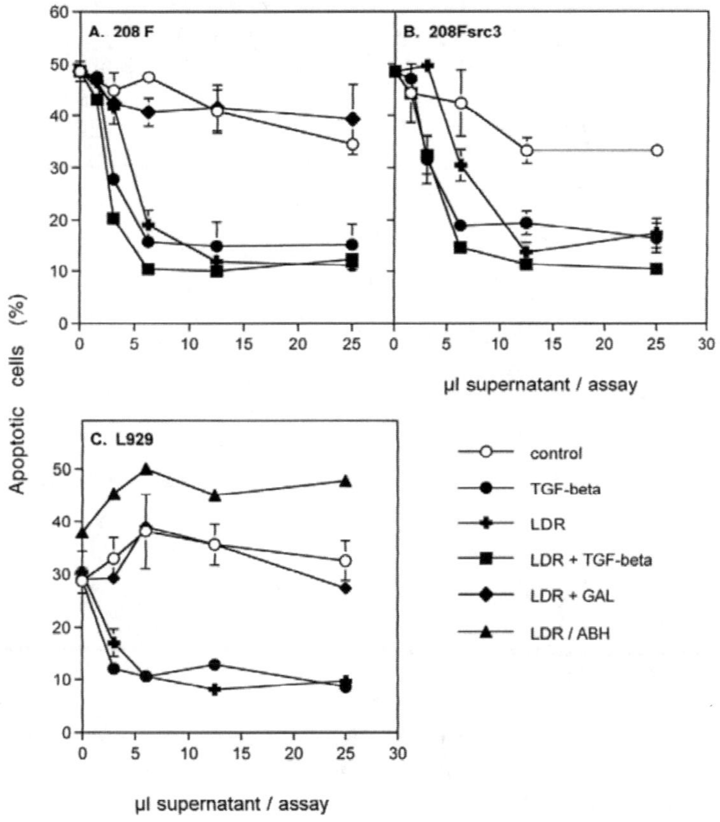

Figure 2. Low dose radiation triggers POD release from nontransformed and transformed cells.

For the evaluation of the significance of radiation-induced effector cell stimulation, a coculture system between transformed cell clumps and dispersely seeded effector cells was used. In this system, a clump of 2000 transformed cells at high local density causes an optimal steady state local concentration of hydrogen peroxide, but the peroxidase concentration is too low for execution of the

Journal of Physics: Conference Series **261** (2011) 012001 doi:10.1088/1742-6596/261/1/012001

HOCl signaling pathway. When normal or transformed effector cells at high number, but low local density, are added to the clumps, these cells compensate for peroxidase but do not undergo apoptosis induction initially, as they show no or only suboptimal hydrogen peroxide production. As shown in figure 3, a clump of src oncogene-transformed cells did not undergo apoptosis even in the presence of TGF-beta. When dispersely seeded nontransformed (figure 3A) or transformed (figure 3B) effector cells had been irradiated before addition the target cell clumps, a dose-dependent increase in apoptosis induction of the transformed target cells was observed. This effect seemed to depend on establishment of intercellular HOCl signaling, as it was efficiently blocked by the peroxidase inhibitor aminobenzyol hydroazide (ABH.)

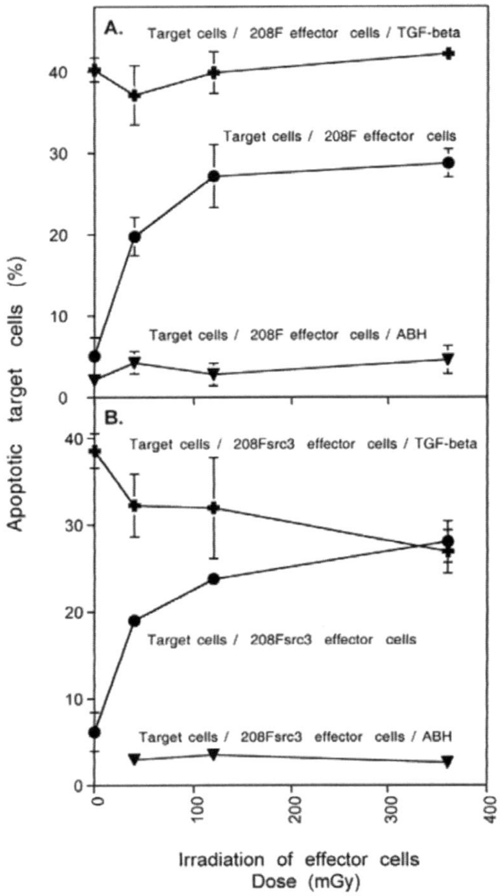

Figure 3. Low dose radiation enhances the effector function
2000 transformed 208Fsrc3 cells in 5µl of medium were spotted as cell clumps into 24 well tissue culture clusters. The cells were incubated at 37 °C in a humidified incubator for about one hour and were overlaid with 0.5 ml complete medium. Nontransformed 208F cells (A) or transformed 208Fsrc3 cells (B) had been irradiated with the indicated low doses of gamma irradition or had remained irradiated. These cells served as effector cells and were seeded to the clumps of unirradiated transformed cells (8000 effector cells per assay). After the effector cells had attached, they received either no addition, 20 ng/ml TGF-beta or 150 µM of the peroxidase inhibitor ABH. After 21 hours (A) or 28 hours (B) the percentage of apoptotic target cells was determined in duplicate assays.

In order to test for the impact of low dose radiation on extracellular superoxide anion generation, inhibition of superoxide anion-dependent apoptosis induction by Cu/Zn-SOD was used for quantiation. Inhibition by Cu/Zn-SOD usually leads to a bell-shaped inhibition curve that allows the quantiation of relative increases or decreases in superoxide anions. Nontransformed cells, both unirradiated or irradiated, did not show significant extracellular superoxide anion generation, as they did not drive peroxynitrite-dependent apoptosis after addition of an NO donor (data not shown). In contrast, transformed 208Fsrc3 cells responded to the fast decaying NO donor DEA NONOate by efficient (peroxynitrite-mediated) apoptosis induction (figure 4). This reaction was blocked by an SOD optimum of 9 U/ml. When half of the standard number of cells had been applied, and therefore half the concentration of superoxide anions was in the system, the SOD concentration for maximal inhibition was only half that of the standard test, indicating that the test system was suitable to monitor relative changes in superoxide anion concentrations. The observed bell-shaped curve allowed easy comparison of the tests. When the 208Fsrc3 cells had been irradiated with 40 mGy of gamma radiation before the test, an eightfold higher concentration of SOD was needed for maximal inhibition, indicating an eightfold enhancement of extracellular superoxide anion production by low dose radiation. Similarly, the inhibition of autocrine apoptosis induction in the gastric carcinoma cell line MKN-45 in the presence of the catalase inhibitor 3-aminotriazole (3-AT) was blocked by SOD in an characteristic way. Low dose irradiated tumor cells required a 32 fold higher concentration of SOD for maximal inhibition of apoptosis. This finding demonstrates the strong impact of low dose radiation on extracellular superoxide anion production by tumor cells.

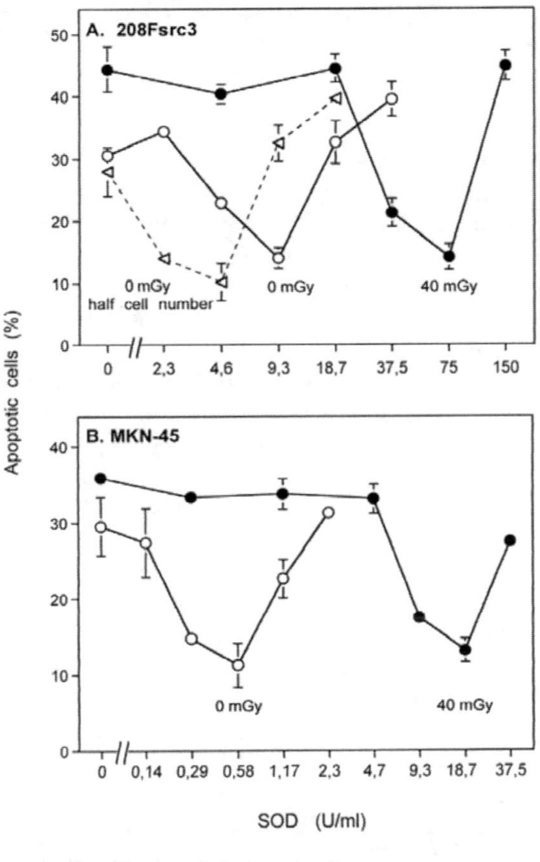

Figure 4. Low dose radiation enhances extracellular superoxide generation in transformed and tumor cells. Transformed 208Fsrc3 cells or human gastric carcinoma cells MKN-45 were irradiated with 40 mGy gamma irradiation or remained untreated.

Cells were seeded in 96 well tissue culture clusters (12 500 cells per well, 100 µl medium). 208Fsrc3 cells received 0.5 mM of the NO donor DEA NONOate for the induction of NO/peroxynitrite-mediated apoptosis induction, in the presence of 20 ng/ml TGF-beta. MKN-45 cells received 150 mM of the catalase inhibitor 3-AT for induction of autocrine ROS-mediated apoptotic selfdestruction. The indicated concentrations of Cu/Zn-SOD were added and the percentages of apoptotic cells were determined after 3.5 hrs (A) or 4 hrs (B) hours. Under (A), half amount of cells was used in a parallel experiment for demonstration of the quantitation of superoxide anion concentration. Additional calibration experiments confirmed a strict linearity between the concentration of superoxide anions and the concentration of SOD necessary for maximal inhibition (data not shown). Control experiments, using the peroxynitrite decomposition catalayst FeTPPS (25 µM) ensured that the reaction used in experiment A was dependent on peroxynitrite formation, control experiments under (B) confirmed autocrine ROS-mediated apoptosis induction through the use of inhibitors of the HOCl pathway [18] (data not shown). 208Fsrc3 cells in the absence of the NO donor and MKN-45 cell in the absence of the catalase inhibitor showed less than 5 % of apoptotic cells at all SOD concentrations (data not shown).

To test for the consequence of low dose radiation-enhanced superoxide anion production by transformed cells, src-transformed cells remained unirradiated or were irradiated with 40 mGy of gamma irradiation. The cells were then placed in small clumps with increasing cell number (figure 5). After the cells had attached, medium in excess and myeloperoxidase (MPO) were added. For control purposes, some of the assays received the HOCl scavenger taurine. As shown in figure 5, MPO-dependent establishment of HOCl signaling and apoptosis induction increased with the density of the clumps in unirradiated target cell assays. 500 cells per clump were too low to allow a detectable apoptosis induction. When the target cells had been irradiated before the experiment started, a strong enhancing effect, especially in the previously suboptimal cell density range was observed. This result is well explained by increased superoxide anion generation by the irradiated transformed cells and subsequent increase in hydrogen peroxide generation, the limiting substrate of HOCl synthesis by MPO. When clumps of irradiated src-transformed cells, again at increasing cell densities, were overlayed with nontransformed dispersely seeded 208F effector cells (figure 5B), a reaction analogous to that shown for MPO (figure 5A) was seen. TGF-beta-treated nontransformed cells induced apoptosis in transformed target cells, thereby showing a stringent requirement for sufficient cell density of the target cells. When the target cells had been irradiated before the beginning of coculture, an increased apoptosis induction, indicating an enhanced contribution of the target cells to intercellular signaling, was seen. The specificities of the signaling reactions were confirmed by inhibition mediated by the HOCl scavenger taurine.

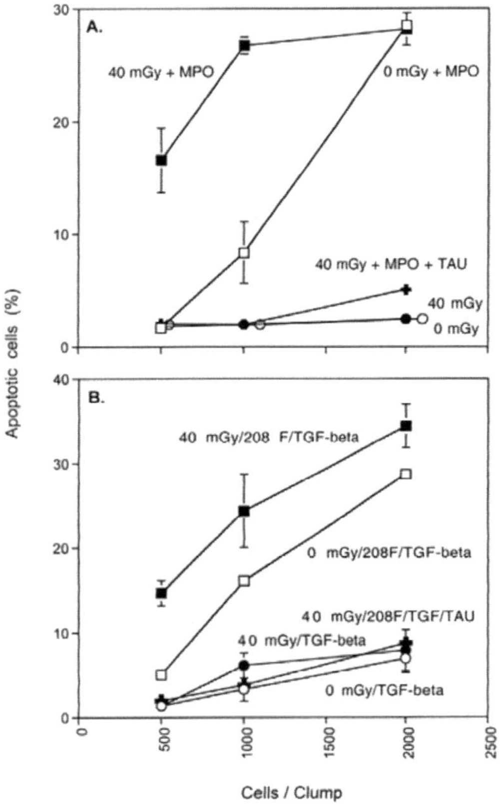

Figure 5. Low dose radiation enhances the target cell function.
Transformed 208Fsrc3 cells were irradiated with 40mGy of gamma irradiation or remained unirradiated, and were then seeded as small clumps of the indicated cell number per 5 µl of medium in 24 well tissue culture clusters. After the cells were attached (within less than one hour), 0.5 ml of medium, containing either no addition ("0 mGy", "40 mGy") or 200 mU/ml myeloperoxidase (MPO)("0 mGy + MPO"; "40 mGy + MPO") were added. In addition, cells irradiated with 40 mGy received 200 mU/ml MPO plus 50 mM taurine (A). In experiment B, 8000 nontransformed 208F cells were added to irradiated or unirradiated transformed cells and the assays received 20 ng/ml TGF-beta or TGF-beta plus 50 mM of the HOCl scavenger taurine, as indicated. Control assays did not receive 208F cells, but TGF-beta. After 31 hours (A) or 28 hours (B) the percentages of apoptotic cells were determined in duplicate assays.

The experiments shown so far demonstrate that low dose gamma radiation strongly enhances POD release from nontransformed, transformed and tumor cells. It also positively modulates superoxide anion generation by transformed and tumor cells, but not by nontransformed cells. There was no effect of low dose radiation on NO release detectable by our test system (data not shown). The experiments shown in figures 3 and 5 demonstrate that removal of a relative small number of transformed cells by neighbouring nontransformed cells is enhanced, when the system is irradiated. Our next question aimed to elucidate the effect of low dose irradiation on a homogeneous population of transformed or tumor cells under conditions, where autocrine apoptosis induction should be possible in principle. Low dose irradiation of transformed 208Fsrc3 cells caused an enhancement of autocrine apoptosis, when the cells had been seeded at suboptimal cell density, but inhibition, when the optimal density had been

Journal of Physics: Conference Series **261** (2011) 012001 doi:10.1088/1742-6596/261/1/012001

used (data not shown). When the gastric carcinoma cell line MKN-45 was subjected to low dose radiation and then brought to conditions suitable for autocrine apoptosis induction, low dose radiation alone had no apoptosis–inducing effect (figure 6). When, in addition, the protective catalase of the tumor cells was inhibited by the catalase inhibitor 3-AT, low dose radiation enhanced apoptosis induction at suboptimal 3-AT concentrations, but inhibited apoptosis at 3-AT concentrations that had been found to be optimal in the absence of low dose radiation treatment. This shift of the optimum curve due to the action of low dose radiation was reversed when a very small concentration of the catalase mimetic EUK-134 removed excess hydrogen peroxide. The analysis of autocrine signaling pathways in MKN-45 cells, without and after low dose radiation revealed that there was not only a shift with respect to the 3-AT concentration, but also a qualitative change in signaling (figure 6). In accordance with previously published work [18], unirradiated MKN-45 cells showed NO/peroxynitrite signaling at low 3-AT concentrations and then a gradual relative increase in HOCl signaling at higher 3-AT concentrations.

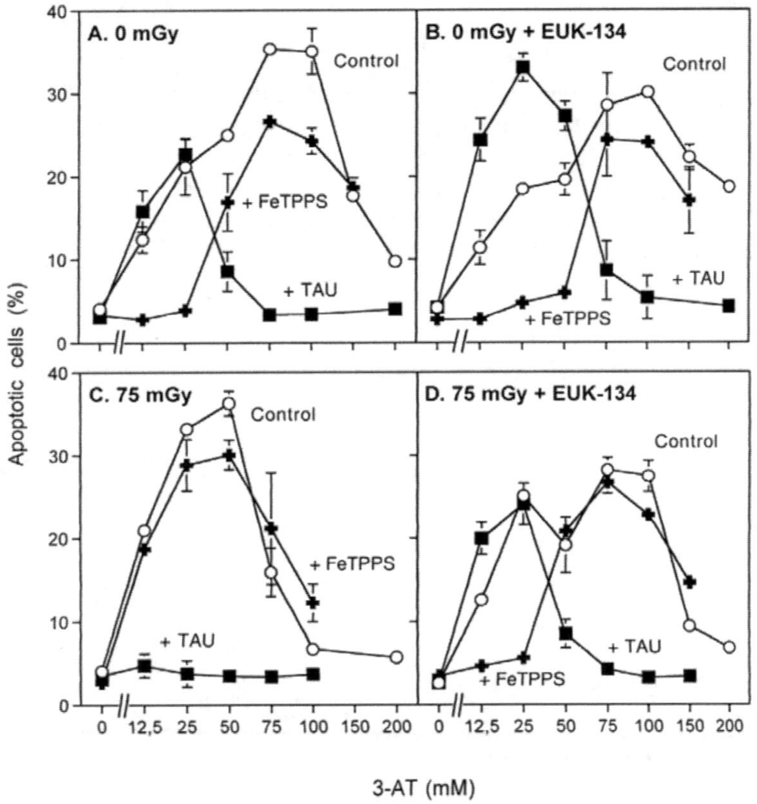

Figure 6. Low dose radiation affects autocrine signaling in tumor cells after catalase inhibition. Gastric carcinoma cells MKN-45 were irradiated with either 75 mGy of gamma irradiation or remained untreated. After irradiation, the cells were incubated for 1 hour at 37 °C and then centrifuged, washed and resuspended in fresh medium. 12 500 cells per 100 µl medium were placed in 96 well tissue culture clusters and received the indicated concentrations of the catalase inhibitior 3-AT, both in the absence or presence of 0.5 µM of the catalase mimetic EUK-134. Some assays were treated in addition with 50 mM of the HOCl scavenger taurine or 25 µM of the peroxynitrite decomposition catalyst FeTPPS. The percentages of apoptotic cells were determined after 3 hours. Note, that in the absence of catalase inhibitor, low dose radiation had not apoptosis-inducing effect on the tumor cells.

The same signaling was observed when small concentrations of EUK-134 had been added to nonirradiated cells. Removal of excess hydrogen peroxide partially reversed the supraoptimal inhibition at very high 3-AT concentrations. When MKN-45 cells had been irradidated at low dose before the experiment, HOCl signaling was dominant at all 3-AT concentrations, without any indication of NO/peroxynitrite signaling. After addition of 0.5 μM EUK-8, the pattern of signaling pathways was completely reversed to the situation of the unirradiated cells, together with a backshift of the optimum curve. These findings indicate that low dose gamma radiation a) has a stimulating effect on suboptimal signaling via the HOCl pathway; b) causes inhibition of the NO/peroxynitrite pathway, which can be replaced by HOCl signaling, and c) causes inhibition of optimal HOCl signaling, especially under conditions of autocrine apoptosis induction.

4. Discussion

This study determines the effect of low dose radiation on intercellular signaling chemistry exerted by nontransformed, transformed and tumor cells. Without irradiation, the nontransformed cells contribute to intercellular ROS signaling (figures 7, 8) through release of effector molecules like POD or NO (please see figure 1 for details), but show no target cell function due to lack of extracellular superoxide anion generation. Transformed cells contribute both by the release of effector molecules and extracellular superoxide anion generation. They thus drive the efficiency and selectivity of intercellular ROS-mediated as well as autocrine apoptosis induction. Tumor cells exhibit target and effector cell functions but block intercellular signaling through interference by a membrane-associated catalase that removes hydrogen peroxide, destroys peroxynitrite and oxidizes NO [18]. Tumor cells can be converted back experimentally to the state of transformed cells when their catalase is inhibited. This experimentel approach allows to study the specific signaling chemistry of tumor cells after abrogation of their catalase-mediated protection.

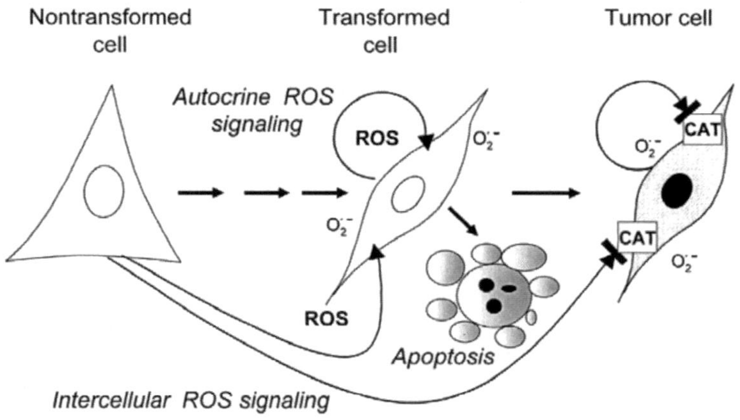

Figure 7. Multistep carcinogenesis and intercellular ROS signaling without irradiation. Nontransformed cells require several distinct steps to reach the transformed state [15,18,53]. Transformed cells generate extracellular superoxide anions that drive the efficiency and selectivity of intercellular ROS signaling between nontransformed effector cells and transformed target cells (please see signaling details in figure 1) as well as autocrine ROS-mediated signaling. This causes apoptosis selectively in transformed cells and may represent a natural control step in oncogenesis. TGF-beta is a central mediator to trigger peroxidase release (not included into the schematics). During tumor progression, tumor cells establish resistance against intercellular and autocrine apoptotic signaling through expression of membrane-associated catalase [16,18].

Journal of Physics: Conference Series **261** (2011) 012001 doi:10.1088/1742-6596/261/1/012001

Low dose radiation-induces peroxidase release from nontransformed, transformed and tumor cells with comparable efficiency (figure 8). In this figure, low dose gamma irradiation leads to the release of peroxidase (POD) from nontransformed (left), transformed (middle) and tumor cells (right) with similar efficiency. Low dose gamma irradiation causes a 8- 32 fold increase in extracellular superoxide anion generation in transformed and tumor cells, but not in nontransformed cells. There is no indication on an increase in NO release from our experiments (data not shown). The increase of signaling components may have different effects on apoptosis induction, dependent on density and nature of the cells. Nontransformed cells remain unaffected, as they do not generate extracellular superoxide anions that are necessary for apoptotic signaling. In nontransformed cells, the increase in superoxide anion and peroxidase production may enhance apoptosis (+), especially under previously suboptimal conditions for apoptosis induction (e.g. low target cell density). The increase in superoxide anion generation by transformed cells may have a negative effect on intercellular induction of apoptosis when the resultant excess hydrogen peroxide (after dismutation of superoxide anions) consumes HOCl (please see figure 1 for details of this consumption reaction). Also, intercellular signaling based on NO/peroxynitrite signaling is inhibited by excess hydrogen peroxide. The increase of superoxide anion generation by tumor cells does not lead to apoptosis induction per se, as the resultant potential increase in ROS signaling is efficiently counteracted by tumor cell membrane-associated catalase. Therefore, low dose radiation affects primarily transformed (premalignant) cells, however in different possible ways. The effects of low dose radiation on tumor cells require parallel modulation of catalase activity, which bring the tumor cells back to the signaling stage of transformed cells.

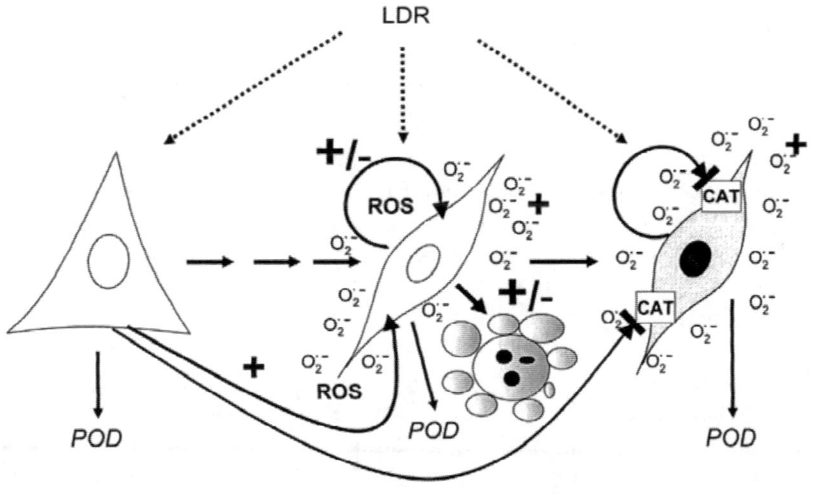

Figure 8. Low dose gamma radiation enhances defined signaling components of intercellular reactive oxygen-mediated apoptosis induction

This activity of low dose radiation resembled that of TGF-beta. As the release of peroxidase was inhibited by galardin, an inhibitor of matrix metalloproteases, peroxidase seems to require the action of proteases in order to be set free. The release of peroxidase by irradiated cells had direct consequences for neighbouring transformed cells in which apoptosis was induced. Therefore, the release of POD by irradiated cells explains the effects reported by Portess et al. [30] Ongoing work defines TGF-beta as a central modulator during radiation induced release of POD and the peroxidase domain of duox as the responsible enzyme (Heinzelmann et al., in preparation). Peroxidase release is necessary for the effector cell action of nontransformed against transformed target cells. For establishment of autocrine apoptosis induction within a population of transformed target cells, membrane-associated peroxidase, not released by matrix metalloprotease can efficiently exert HOCl signaling, but has a negative effect on NO/peroxynitrite signaling due to a destructive effect on peroxynitrite due to high local concentrations of the peroxidase (Bauer, unpublished result).

Low dose radiation also has a marked effect on extracellular superoxide anion production by transformed and tumor cells, whereas nontransformed cells do not show such a response. An ongoing study, using nontransformed cells with an inducible *RAS* oncogene shows that only cells that already express *RAS* and generate superoxide anions at a measurable extend are enhanced in their superoxide anion generation by low dose radiation. Therefore, the enhancement of the target cell function superoxide anion release maintains the selectivity of this feature with respect to the malignant phenotype. An obvious consequence of the enhancement of superoxide anion generation is an enhancement in intercellular and autocrine apoptosis induction under conditions that are suboptimal for apoptosis induction without radiation. Such situations are given when a low number or low local density of target cells is present. Under these conditions, a clear positive effect of low dose radiation on the removal of these cells is observed in vitro and can be predicted for the situation in vivo. Thereby, the enhancement of superoxide anion production and POD release can be expected to interact in a synergistic way, establishing optimal HOCl signaling. However, our data also show that opposite effects can be induced by low dose radiation. Irradiation of cells at substantial density and optimal conditions may lead to an excess of hydrogen peroxide (derived from stimulated superoxide anion production) in relationship to available POD. Under these conditions, the consumption reaction between HOCl and hydrogen peroxide may cause a massive or even complete removal of HOCl and thus inhibit the HOCl signaling pathway. In this situation, low dose radiation would have a protective rather than a destructive effect on malignant cells. These findings are in agreement with a recent discussion of the potential impact of low dose radiation on intercellular signaling [15]. The study of catalase-inhibited MKN 45 cells (that had been experimentally brought back to the stage of transformed cells) showed that low dose radiation not only shifted the optimum curve with respect to catalase inhibition, but that it also changed the quality of signaling. Low dose radiation caused the disappearance of NO/peroxynitrite signaling that was observed in unirradiated tumor cells at low concentrations of the catalase inhibitor. The negative effect on NO/peroxynitrite signaling is explained by the increased consumption of NO in the presence of an enhanced hydrogen peroxide concentration and by the obvious lack of low dose radiation to enhance NO generation in a detectable way. Therefore it was not unexpected that signaling in a catalase-inhibited Ewing sarcoma cell line, that was completely and solely dependent on NO/peroxynitrite signaling, was inhibited by low dose radiation to a significant degree (data not shown).

Taken together, our experimental data provide rational explanations for the effects obseverved in a previous paper [30] and demonstrate the potential of low dose radiation to remove malignant cells in the course of tumor establishment and progression. They thus might represent the basis for the protective low dose radiation effects observed in vitro and in vivo [32-44]. However, it is also shown in this paper that cells with certain specific signaling features, ratios between specific signaling components as well as certain situations with respect to the density of cells may lead to protection of malignant cells rather than to their destruction. Which of the two alternative effects dominates *in vivo* is not predictable from the data obtained *in vitro*. Therefore, we suggest not to use the term "protective apoptosis mechanism (PAM)" as synonym for intercellular induction of apoptosis, as this would imply that intercellular ROS signaling always favours removal of malignant cells. Our data indicate that this must not necessarily be always the case. These aspects require further experimentation. It will be a challenge to find out, whether a signaling phenotype restriced to NO / PON signaling is advantageous to transformed cells that exhibit this phenotype, as these cells might be protected from low dose radiation effects through negative effects of increased hydrogen peroxide on NO/peroxynitrite signaling. The specific NO/peroxynitrite signaling phenotype might represent a phenotype that is associated with the potential to escape control by low dose radiation, even in the absence of protective catalase.

Autocrine and intercellular ROS signaling is abrogated in tumor cells that have established protection against intercellular signaling through catalase expression at the outer surface of the tumor cells. Low dose radiation alone does not abrogate resistance, as seen in figure 6. Therefore, the concentration of superoxide anions and hydrogen peroxide reached after low dose radiation seems not to be in the concentration range that is necessary to overrun protective catalase. In a recent paper [16], it had been shown that excess hydrogen peroxide generated by added glucose oxidase (GOX) can re-establish intercellular ROS signaling. The effect of GOX was markedly enhanced by exogenous MPO. Interestingly and in line with these published observations, addition of MPO to low dose irradiated, but not to unirradiated tumor cells caused strong apoptosis induction (data not shown). This finding indicates the potential of low dose radiation to enhance apoptosis induction in tumor cells, when parallel signaling requirements are also provided. We are presently establishing the signaling conditions that, predicted by the detailed knowledge of intercellular ROS signaling, should allow to define situations under which protective catalase of tumor cells is overrun, inactivated or even destroyed by the concerted action of specific signaling components. This rational approach may be instrumental to utilize low dose radiation effects for specific therapeutic applications in the future.

Acknowledgements
This work was supported by a grant from EC RiscRad-contract-FI6R-CT-2003-508842. I am grateful for valuable intellectual support by the COST consortium 'ChemBioRadical' (COST Action CM0603). I thank Dr. P. Kundrát (Helmholtz Zentrum München) for critical comments on the manuscript.

References

[1] Irani K, Xia Y, Zweier JL, Sollott SJ, Der CJ, Fearon ER, Sundaresan M, Finkel T and Goldschmidt-Clermont PJ 1997 *Science* **275** 1649.
[2] Suh Y-A, Arnold RS, Lassegue B, Shi J, Xu X, Sorescu D, Chung AB, Griendling KK and Lambeth JD 1999. *Nature* **401** 79.
[3] Yang JQ, Li S, Domann FE, Buettner G and Oberley LW 1999 *Mol. Carcinogen.* **26** 180.
[4] Heigold S, Sers C, Bechtel W, Ivanovas B, Schäfer R and Bauer G 2002. *Carcinogenesis* **23** 929.
[5] Mitsushita J, Lambeth JD and Kamata T 2004 *Cancer Res.* **64** 3580.
[6] Arnold RS, Shi J, Murad E, Whalen AM, Sun .Q, Palavarapu R, Parthasarathy S, Petros JA and Lambeth JD 2001. *Proc. Natl. Acad. Sci. USA* **98**, 5550.
[7] Tominaga K, Kawahara T, Sano T, Toida K, Kuwano Y, Sasaki H and Kawai T 2007 *Free Radical Biol. & Med.* **43** 1627.
[8] Laurent E, McCoy JW, Maccina RA, Liu W, Cheng GJ, Robine S, Papkoff J and Lambeth JD 2008 *Int. J. Cancer* **123** 100.
[9] Jürgensmeier J, Schmitt CP, Viesel E, Höfler P and Bauer G 1994 *Cancer Res.* **54** 393.
[10] Beck E, Schäfer R and Bauer G 1997. *Exp. Cell Res.* **234** 47.
[11] Herdener M, Heigold S, Saran M. and Bauer G 2000 *Free Radical Biol. Med.* **29** 1260.
[12] Schwieger A, Bauer L, Hanusch J, Sers C, Schäfer R and Bauer G 2001 Carcinogenesis *22* 1385.
[13] Bauer G. 2000. *Anticancer Res.* **20** 4115.
[14] Bauer G 2002. *Prostaglandines, Leukotrienes and Essential Fatty Acids* **66** 41.
[15] Bauer G 2007. *Int. J. Radiat. Biol.*. **83** 887.
[16] Bechtel W and Bauer G 2009 *Anticancer Res.* **29** 4541.
[17] Bechtel W and Bauer G 2009 *Anticancer Res* **29** 4559.
[18] Heinzelmann S and Bauer G 2010. *Biol. Chem.* **391** 675.
[19] Steinebach C and Bauer G 2001 *In vitro Appl.Mol. Toxicol.* **14** 107.
[20] Schimmel M. and Bauer G.2002. *Oncogene* **21** 5886.
[21] Engelmann I, Eichholtz-Wirth H and Bauer G 2000 *Anticancer Res.* **20** 2361.

[22] Deichman G, Matveeva VA, Kashkina LM, Dyakova NA, Uvarova EN, Nikiforov MA and Gudkov AV 1998 *Int. J. Cancer* **75** 277.
[23] Deichman G 2000 *Biochemistry (Moscow)* **65** 78.
[24] Barcellos-Hoff MH and Ravani SA. 2000 *Cancer Res.* **60** 1254.
[25] Barcellos-Hoff MH 2001 *J. Mammary Gland Biol. Neoplasia* **6** 213.
[26] Barcellos-Hoff MH and Brooks AL 2001 *Radiat. Res.* **156** 618.
[27] Bauer G 1996 *Histol. Histopathol.* **11** 237.
[28] Barcellos-Hoff MH 2002 *Radiat. Res.* **158** 792.
[29] Barcellos-Hoff MH and Costes SV 2006 *Mut. Res.* **597** 32.
[30] Portess DI, Bauer G, Hill MA and O`Neill P 2007 *Cancer Res.* **67** 1246.
[31] Tubiana M. 2009 *Cancer Radiotherapie* **13** 238.
[32] Averbeck D 2010 *Mutation Res. – Fund. Mol.Mechanism Mutagen.* **68 7**.
[33] Azzam EI, Raaphorst GP and Mitchel RE 1994 *Radiat. Res.* **138** S28.
[34] Azzam EI, de Toledo SM, Raaphorst GP and Mitchel RE 1996 *Radiat. Res.* **146** 369.
[35] Elmore E, Lao XY, Kapadia R and Redpath JL 2006 *Radiat. Res.* **166** 832.
[36] Ko M, Lao XY, Kapadia R, Elmore E and Redpath JL 2006 *Mut. Res.* **597** 11.
[37] Mitchel REJ, Jackson JS, Morrison DP and Arlisle SM 2003 *Radiat. Res.* **162** 20.
[38] Mitchel REJ 2007 *Dose Response* **5** 1.
[39] Redpath JL 2005 *Nonlinearity Biol. Toxicol. Med.* **3** 113.
[40] Redpath JL 2006 *Dose-Response* **4** 302.
[41] Redpath JL, Liang D, Taylor TH, Christie C and Elmore E 2001 *Radiat. Res.* **156** 700.
[42] Redpath JL, Lu Q, Lao X, Molloi S and Elmore E 2003 *Int. J. Radiat. Biol.* **79** 235.
[43] Pant MC, Liao XY, Lu O, Molloi S, Elmore E and Redpath JL 2003 *Carcinogenesis* **24**1961.
[44] Yu H, Song A, LU Y, Iu W and Shen F 2004 *Chin. Med. J.* (English Edition) **117** 1036.
[45] Kadhim MA 2003 *Oncogene* **22** 6994.
[46] Lyng FM, Maquire P, Kilmurray N, Mothersil C, Shao C, Folkard M and Prise K 2006 *Int. J. Radiat. Biol.* **82** 393.
[47] Schoellnberger H and Eckl PM 2007 *Dose-Response* **5** 187.
[48] Schoellnberger H, Mitchel RE, Azzam EI, Crawford-Brown DJ and Hofmann W 2002. *Int. J. Radiat. Biol.* **78** 1159.
[49] Scott BR 2007 *Dose response* **5** 131.
[50] Scott BR, Walker DM, Tesfaigzi Y, Schöllnberger H and Walker V 2003 *Nonlinearity Biol. Toxicol. Med.* **193** 122.
[51] Scott BR 2004 *Mutation Res.-Fund. Mol. Mechanism Mutagen.* **568** 129.
[52] Haberstroh K, Heigold S and Bauer G 2002 *Int J Oncol.* **21** 145.
[53] Hanahan D and Weinberg RA 2000 *Cell* **100** 57.

Journal of Physics: Conference Series **261** (2011) 012002 doi:10.1088/1742-6596/261/1/012002

UVA-induced damage to DNA and proteins: direct *versus* indirect photochemical processes

P M Girard[1,2,3], S Francesconi[1,2,3], M Pozzebon[1,2,3], D Graindorge[1,2,3], P Rochette[4], R Drouin[5] and E Sage[1,2,3]

[1] Institut Curie, Bât. 110, Centre Universitaire, 91405 Orsay cedex (France), [2] CNRS UMR3348, Bât. 110, Centre Universitaire, 91405 Orsay cedex (France), [3] University of Paris-sud XI, 91405 Orsay cedex (France), [4] Génie tissulaire et régénération:LOEX, Centre de recherche FRSQ du Centre hospitalier affilié universitaire de Québec, Laval University, Quebec City, Québec (Canada), [5] Division of Genetics, Department of Pediatrics, Faculty of Medicine and Health Sciences, Université de Sherbrooke, Sherbrooke, Québec (Canada)

E-mail: evelyne.sage@curie.fr

Abstract. UVA has long been known for generating an oxidative stress in cells. In this paper we review the different types of DNA damage induced by UVA, i.e. strand breaks, bipyrimidine photoproducts, and oxidatively damaged bases. Emphasis is given to the mechanism of formation that is further illustrated by the presentation of new *in vitro* data. Examples of oxidation of proteins involved in DNA metabolism are also given.

1. Introduction

Even though there is a continuum of energies in the electromagnetic spectrum of radiations, the UV portion of sunlight has been divided into UVC (below 280 nm), UVB (280-315 nm) and UVA (315-400 nm) for categorizing the effects of solar exposure to human skin. Wavelengths below 290 nm are blocked by the stratospheric ozone. These wavelengths are of the highest energies (> 4.4 eV), coincide with maximal absorption by DNA and other biomolecules, and would be extremely damaging to genetic material and biological tissues in general. Therefore the stratospheric ozone layer protects life from UVC radiation. Longer wavelength UVB (λ > approximately 290 nm) penetrates the ozone layer and constitutes about 5-10% of the terrestrial solar UV radiation. It is absorbed by nucleic acids and has the ability to directly cause genotoxic damage to DNA and ultimately skin cancer.

In contrast, wavelengths in the UVA range received little attention up to the nineties, although they are by far the most abundant solar UV radiation (above 90%) that reaches the surface of earth. UVA radiation also penetrates human skin more effectively than UVB. Unlike UVB, the UVA component of solar radiation is weakly absorbed by DNA [1], but rather excites other endogenous chromophores, generating various reactive oxygen species (ROS) in cells. Moreover, most UVA-mediated biological events are oxygen-dependent. These observations provided the basis for considering UVA as a strong generator of oxidative stress. Singlet oxygen is the major ROS involved in UVA-mediated cell inactivation [2] and gene activation [3], it is also a primary source of UVA-mediated signalling [4]. Once thought to be relatively innocuous, UVA is now known as a damaging agent for DNA, proteins, lipids, with harmful consequences such as skin ageing and carcinogenesis. Due to the popularity of the UVA tanning saloons and to the widespread use of efficient UVB-absorbing sunscreens blocking erythema, while accompanied by prolonged periods of sunbathing, exposure of the human population to UVA has increased significantly in the last decades. As a consequence, the deleterious effect of UVA has recently emerged as a source of concern for public health. As an example, sunbeds have been included into the list of carcinogens to humans by the International Agency of Research on Cancer [5]. The cellular and molecular responses to UVA in relation to skin carcinogenesis have been recently reviewed [6]. Here, we focus on damage induced to DNA and protein by UVA radiation and discuss the possible mechanisms of formation.

2. DNA damage induced by UVA

As stated in the introduction, UVC and UVB radiation are directly absorbed by DNA, causing the formation of cyclobutane pyrimidine dimers (CPDs), pyrimidine (6-4) pyrimidone photoproducts (6-4PPs) and their Dewar valence isomers. These bipyrimidine photoproducts are not expected to be formed by UVA because of the inability of DNA to readily absorb in the UVA range. Meanwhile, a pioneer work of Tyrrell's group reported the induction of CPDs by very pure narrow-band 365-nm UVA in bacteria, with a higher relative frequency of TT CPDs and a reduced excision of these CPDs, comparatively to 254-nm UVC [7,8]. The same group also reported the induction of single strand breaks (alkali-labile bonds) by UVA in bacteria and phage DNA [9]. It is well established that UVA promotes photosensitization of DNA, triggered by singlet oxygen (type II photosensitization), electron abstraction (type I photosensitization) or hydroxyl radicals. Therefore, the formation of oxidative DNA damage is expected. The induction of single strand beaks (SSB) and DNA-protein crosslinks has indeed been observed in mammalian cells upon UVA irradiation [10]. In the nineties, the availability of new and sensitive methods allowed the efficient measurement of various types of DNA lesions. These methods are comet assays, commercially available specific antibodies and DNA repair enzymes, HPLC coupled with electrochemical detection (HPLC-ED) to reveal 8-oxoguanine, and more recently HPLC coupled with tandem mass spectrometry (HPLC-MS/MS) to measure 8-oxoguanine and all the different bipyrimidine photoproducts. UVA-induced DNA damage has thus been timely revisited.

The *Escherichia coli* Fpg or Nth proteins have been used to reveal oxidation products of purine (mainly 8-oxoG) or pyrimidine, respectively, and Nfo protein to reveal abasic sites. The 8-oxoguanine (8-oxoG) has been more specifically measured by HPLC-ED. We and other have shown that UVA radiation induces single strand breaks (SSB), oxidized pyrimidines, oxidized purines in mammalian cells [11-14], and that the most abundant oxidative DNA damage produced is 8-oxoG [15-17].

By the use of sensitive methods, CPDs have been detected in mammalian cells, including in human skin upon UVA irradiation [11, 13, 15, 17, 18-22 and 23 for review]. It has been found that CPDs are produced about 10^5 fold less efficiently by UVA than by 254 nm UVC [7,19]. 6-4 photoproducts were barely detected using specific antibodies [19]. In agreement with the original studies in bacteria [7], the distribution of CPDs induced by UVA drastically differs from that produced by UVC, UVB or simulated sunlight (SSL). UVA-induced CPDs predominantly form at TT sites, while those produced at CC sites are poorly represented [20]. HPLC-MS/MS also revealed predominance for TT CPDs over CT and TC CPDs, whereas CC CPDs and 6-4PPs were not detected [17,22]. As shown in Fig. 1, CPDs, 8-oxoG, oxidatively damaged pyrimidines, and SSB are formed in the ratio 10:3:1:1. Interestingly, CPDs are the major lesions induced by UVA in mammalian cells and human skin [17,22].

Moreover, it has been shown that UVA is able to photoisomerize 6-4 photoproducts produced by the UVB photons of simulated sunlight into Dewar photoproducts [17,19]. This sustains the observation that Dewar photoproducts are more frequently produced by natural and simulated sunlight, than by UVB [19]. DNA double strand breaks, revealed by neutral comet assay or γ-H2AX foci (phosphorylated H2AX variant histone), are not readily produced in human transformed fibroblasts or keratinocytes by direct UVA irradiation, but occurs at low extent during the process of DNA lesions [24,25].

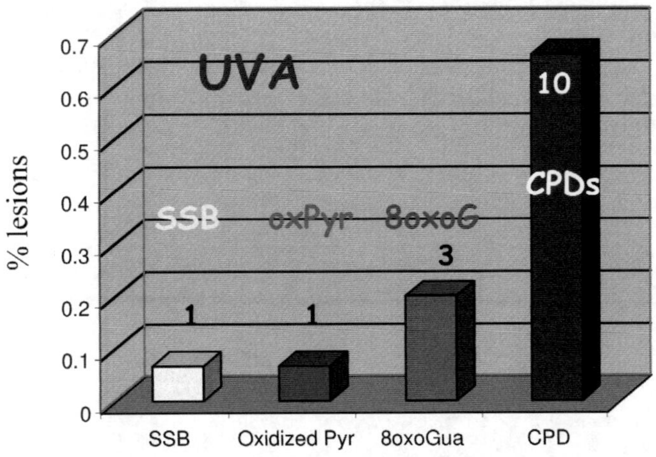

Figure 1. Distribution of CPDs, 8oxoguanine (8oxoG), oxidatively damaged pyrimidines (oxPy), single strand breaks (SSB), revealed respectively by T4 phage DenV protein, bacterial Nth and Fpg proteins in combination with alkaline agarose gel, in Chinese hamster ovary cells exposed to UVA. The yields of formation were obtained by linear regression of the amount of lesions with respect to the dose (from [17]).

3. Processes involved in the formation of UVA-induced DNA damage

The induction of single strand breaks in bacterial DNA by narrow-band 365-nm radiation is totally dependent on the presence of oxygen [9]. The low amounts of SSB and oxidized pyrimidines, likely produced by hydroxyl radicals originated from Fenton reaction, and the predominance for 8-oxoG likely sign a major contribution of singlet oxygen in the formation of oxidative DNA damage in cells by UVA [23]. A type I mechanism by hydrogen or electron abstraction mediated by a photosensitizer cannot be totally excluded.

The photochemical process underlying the formation of UVA-induced CPDs has been a matter of debate for many years. The quantum yields were 7 x 10^5 less at 365 nm than at 254 nm in bacteria [7] and 10^5 less for broad-band UVA than at 254 nm [19]. Such yield of CPDs formation by UVA is in perfect agreement with the DNA absorption spectrum [1]. This led us to conclude that CPDs formation by UVA could be compatible with direct absorption by DNA [19]. However, in bacteria, the longer wavelength radiation induced predominantly thymine dimers (TT:CT in the ratio 5:1), whereas at the shorter wavelength CPDs were induced in the ratio 5:4:1 (TT:CT:CC) [7]. In mammalian cells, CPDs also predominately formed at TT sites at a gene level [20] and at a higher extent in the bulk DNA [17,21,22]. The principal induction of CPDs at T-containing bipyrimidine sites, and mostly at TT sites, is a reminiscence of what was observed by photosensitization reaction of DNA by aromatic ketones [26, 27]. This could suggest that, in contrast to what was stated above, most of CPDs are formed through triplet energy transfer from photosensitizer to excited thymine. The identification of an hypothetical UVA-absorbing chromophore tightly bound to DNA that allows energy transfer remains elusive.

Evidence recently emerged in favour of a direct absorption of UVA by DNA that mediates CPD formation. Firstly, the distribution of the different types of DNA damage induced by UVA is similar in plasmid or isolated DNA and in cells [28-31]. In particular, upon UVA irradiation, CPDs are formed in plasmid DNA by a mechanism that should not involve photosensitization reactions, due to the absence of photosensitizer in the irradiation mixture. Therefore, CPDs in plasmid or isolated DNA should be formed by direct absorption. Moreover, it has clearly been demonstrated that UVA radiation

Journal of Physics: Conference Series **261** (2011) 012002 doi:10.1088/1742-6596/261/1/012002

is able to induce 6-4PPs in plasmid DNA [30], while 6-4PPs are produced in tiny amount in cells [19]. Importantly, 6-4PPs cannot be formed by photosensitization and are only produced by direct absorption. To highlight our point of view, we provide new evidence in favour of a direct mechanism. Figures 2 and 3 show that CPD distributions along the adenylphosphoribosyl transferase (APRT) gene are superimposable whether isolated cellular DNA or cells were UVA-irradiated, as also observed for UVB or UVC-induced CPDs which are formed exclusively by direct absorption.

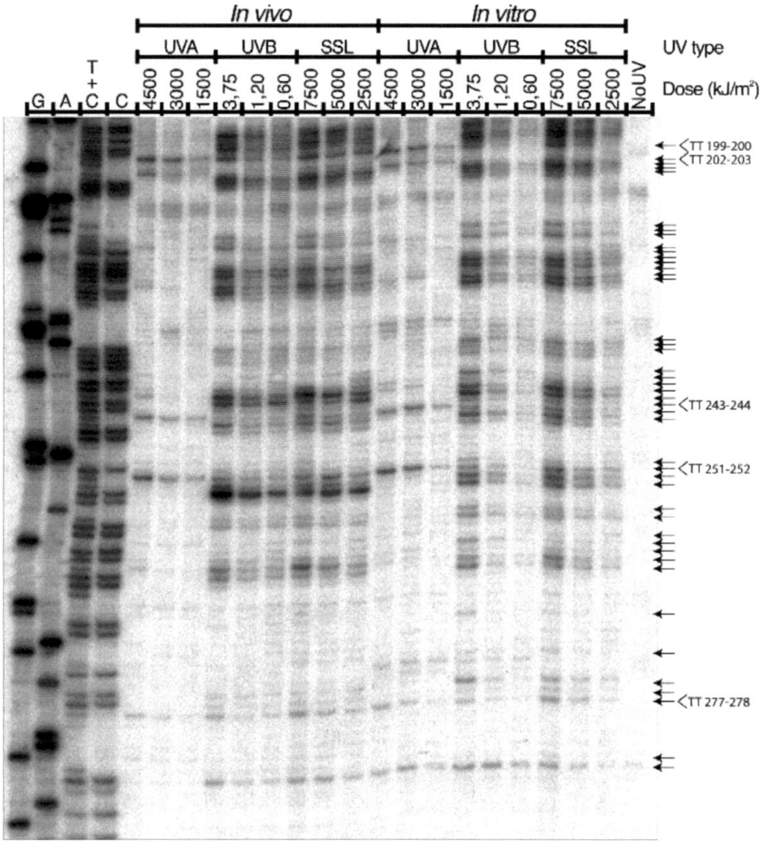

Figure 2. Induction of CPDs by UVA, UVB, and simulated sunlight (SSL), at nucleotide resolution, in the aprt exon 2 gene of Chinese hamster ovary cells. Cells were either unirradiated (No-UV) or irradiated as indicated and CPD formation was analyzed by LMPCR along the non-transcribed strand of exon 2 of aprt gene. The first four lanes from the left of the autoradiogram exhibit LMPCR of DNA treated in standard Maxam-Gilbert cleavage reactions. For the next 18 lanes, each band in a lane corresponds to a site of CPD formation. Either cells (*in cellulo*) or DNA isolated from cells (*in vitro*) were irradiated with 4500, 3000, 1500 kJm^{-2} of UVA, 3.75, 1.20, 0.60 kJm^{-2} of UVB, or 7500, 5000, 2500 kJm^{-2} of SSL, as indicated. The doses for UVB, SSL, and UVC induced approximately equal global CPD. The red dots indicate UVA-induced CPDs at TT sites. The figure reveals that the patterns of bands (sites of CPDs) produced by UVB, SSL and UVA are the same whether DNA isolated from cells or cells were exposed to radiation. Moreover, it shows that the pattern of band for UVA differs from that of UVB or SSL and reveals formation of CPDs at TT sites (compare with Maxam and Gilbert ladder). For further details in procedures see [20].

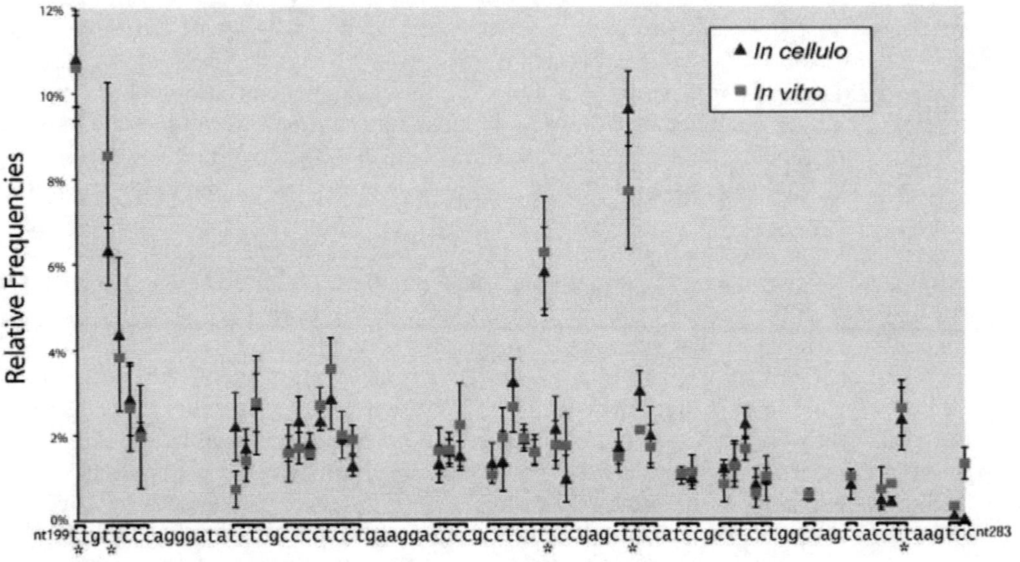

Figure 3. The sites and frequencies of CPDs formation *in vitro* and *in vivo* by UVA are similar. This figure represents the mean of quantified LM-PCR experiments such as those described in figure 2. For further details in quantification see [20]. The highest frequencies for UVA-induced CPDs are at TT sites.

Moreover, the frequencies of CPD formation *in vitro* and *in cellulo* are the same in these experiments, as well as in data presented in Table 1. Finally, an other argument is provided by the similarities in the yields of CPD formation within calf thymus DNA and keratinocytes exposed to UVA through filters exhibiting increasing cut-off wavelength (322-390 nm) [31]. Collectively, these observations strongly suggest that the photochemical process underlying the formation of CPDs by UVA is direct absorption of UVA photons by DNA. The low energy of UVA photons is sufficient to directly excite DNA with subsequent formation of bipyrimidine photoproducts, both CPDs and 6-4PPs, yet at low level. The predominance for CPD formation at TT sites may be explained by the fact that thymine has, in fact, the lowest triplet state energy with regard to other bases. In contrast, 8-oxoG is essentially formed by photosensitization reaction mediated by singlet oxygen.

Table 1. Yields of CPDs formation (per kbp per Jm^{-2}) by broad-band UVA, UVB and UVC in Chinese hamster cells (*cellulo*) [17] and in plasmid DNA (*vitro*) [28].

	in vivo	*in vitro*
UVC (254 nm)	2.2 (+/-0.2) $x10^{-2}$	1.5 (+/-0.3) $x10^{-2}$
UVB (λ>300 nm)	2.0 (+/-0.2) $x10^{-4}$	1.5 (+/-0.4) $x10^{-4}$
UVA	3.0 (+/-0.4) $x10^{-7}$	1.4 (+/-0.2) $x10^{-7}$

4. Protein oxidation induced by UVA

Proteins are major cellular targets for photooxidation. This is due to their abundance in cells and also to the fact that they carry chromophores, *i.e.* primarily side chains or bound photosensitizers. Proteins absorbs in the 250-320 nm region due to the aromatic amino acids. Some proteins bind prosthetic groups which absorb in the UVA and visible range, such as flavins, hemes. As a result, direct photooxidation may occur and arises from absorption of radiation by the protein or by the bound chromophores, therefore generating excited states or radicals through type I processes. Indirect photooxidation of proteins *via* type II mechanism may also occur and involves singlet oxygen generated by energy transfer from either protein-bound or other endogenous chromophores to oxygen. All these processes have been detailed in [32-34]. All amino acid residues are susceptible to oxidation [35]. Anyhow, the thiol function in cystein residues is among the most susceptible oxidant-sensitive targets within proteins and can undergo various reversible and irreversible redox alterations in response to ROS.

UVA, as an oxidative agent, has been shown to produce various protein damage [36]. In fact, the observation that near UV wavelength is able to inactivate photoreactivating enzyme was made more than 30 years ago [37]. More recently, in human skin cells, UVA irradiation was shown to induce sulphydryl oxidation in proteins that depends on iron, singlet oxygen and hydrogen peroxide [38]. UVA also caused changes in catalase charge in epidermis reconstituted with low phototype melanocytes, possibly due to oxidative modification of Trp and Met residues [39]. Photoageing has also been associated with protein oxidation in human skin, with accumulation of protein carbonyls, especially in the upper dermis. It was recently shown that the repair enzyme methionine-*S*-sulfoxide reductase (MSRA) was expressed in human epidermis and specifically upregulated by low doses of UVA radiation [40]. MSRA, by reducing the methionine sulfoxide (MetO) to methionine, thus reverses the inactivation of proteins caused by oxidation of critical methionine residues. Another example is given with the base excision repair protein, DNA *N*-glycosylase hOGG1 which removes 8-oxoguanine and has been shown to be sensitive to oxidation, e.g. cystein oxidation [41]. Interestingly, glutathione depletion of skin fibroblasts and melanoma cells, which increases oxidative stress, causes a strong repair retardation of 8-oxoguanine after exposure to non-toxic doses of visible light or UVA radiation [42]. Moreover, the cleavage activity at 8-oxoguanine residues measured in protein extracts from irradiated cells drops transiently after irradiation. A modification of hOGG1 is suspected, but is not caused by critical oxidation of Cys.

We have recently shown that exposure of eucaryotic cells (*i.e.* fission yeast *S. pombe* and mammalian cells) to oxidative stress induced by UVA radiation led to inhibition of DNA replication by a yet unknown process, which does not require a functional DNA damage checkpoint response, despite $ATM^{(spMec1)}$-, $ATR(^{spRad3)}$-, $p38(^{spSty1)}$-dependent pathways activation [24,43]. Data led us to propose that inhibition of DNA replication is due to impaired replication fork progression, rather as a consequence of UVA-induced oxidative damage to protein than to DNA. To highlight this point, we carried investigations on two proteins involved in DNA metabolism, Proliferating Cell Nuclear Antigen (PCNA), a cofactor of DNA polymerases, and XRCC3 protein involved in repair of double strand breaks by homologous recombination.

PCNA is a homotrimeric protein which acts as sliding clamp on DNA, conferring high processivity to replicative DNA polymerases. It also plays a pivotal role in translesion synthesis. PCNA undergoes posttranslational modification such as mono- or poly- ubiquitylation upon cells treatment with UVC or hydroxyurea which block DNA replication forks. In the left part of figure 4, the bands migrating slower than monomeric PCNA likely reflect ubiquitilated PCNA. Such PCNA modification occurs in active PCNA, bound to chromatin. It is not induced by UVA. In contrast, right part of figure 4 shows that a small fraction of PCNA present in the soluble protein fraction migrates as a trimer (90 kDa protein) and is specifically induced by UVA. This UVA-induced PCNA modification increases in a dose-dependent manner, it occurs very rapidly and it is stable for more than 6 hours after irradiation (not shown). It is largely reduced when UVA irradiation is performed in the presence of the antioxidant N-acetylcysteine or NaN_3, a quencher of singlet oxygen. Altogether, it appears that UVA-induced ROS lead to irreversible, oxidative, covalent cross-linking between the three subunits of PCNA. Such an irreversible PCNA crosslinking has been previously reported [44] upon treatment of human cells with azathioprine plus low doses UVA. Singlet oxygen thus produced by photosensitization was also responsible for this PCNA modification. The authors suspect an interstrand crosslink involving His153 and Lys77, since no covalent trimeric PCNA is observed in cells carrying an amino acid change in PCNA at His153. The chemical process underlying this covalent reaction is still unknown.

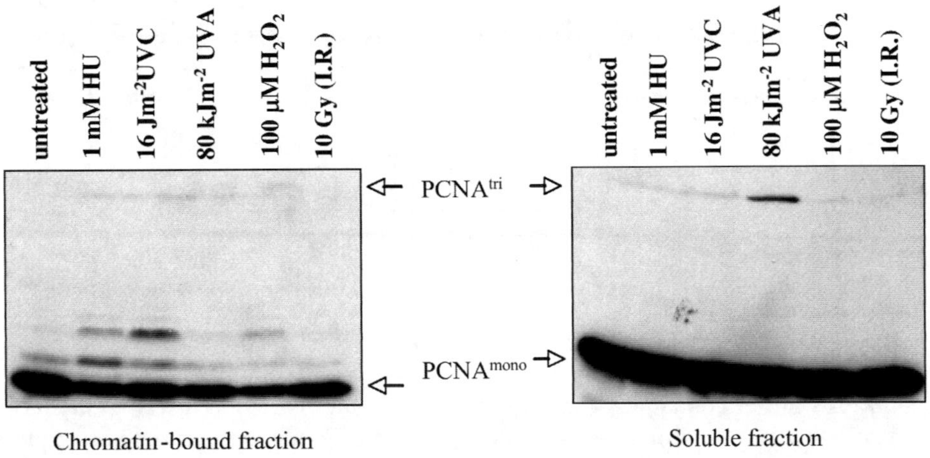

Figure 4. MRC5Vi cells were either untreated or exposed to 1mM hydroxyurea (HU) for 16h, 16 J/m² UVC, 80 kJ/m² UVA, 100 mM H_2O_2 for 30 min or 10 Gy γ-rays. Immediately post treatment, cells were lysed on ice in lysis buffer and total soluble and insoluble protein were loaded on reducing 9 % SDS-PAGE. Proteins were then transferred onto nitrocellulose membrane and PCNA protein detected using mouse monoclonal anti-PCNA antibody (PC10, Santa Cruz Biotechnology). $PCNA^{mono}$ = PCNA monomeric. $PCNA^{tri}$ = PCNA trimeric.

In vertebrates, XRCC3 is one of the five RAD51 paralogs which play non-redundant essential functions in the repair of DNA double strand breaks by homologous recombination. XRCC3 deficiency results in impaired DNA damage-induced RAD51 foci formation, impaired homologous recombination and elevated chromosome aberrations. XRCC3 contains 8 cystein residues, and in normal growing conditions, XRCC3 is mainly reduced (thiol groups). Protein extracts from human cells HCT116 deficient ($XRCC3^{-/-}$) or not in XRCC3 were prepared and migrated on SDS-PAGE. The comparison of the bands (see non-reducing conditions in figure 5) allowed to allocate the fast migrating band to XRCC3 protein (absent in XRCC3 deficient cells). When cells were UVA-irradiated and protein extracts migrated under non-reducing conditions, XRCC3 protein band decreases, to almost disappear at 160 kJm^{-2}, while the protein is still present in the extract as revealed when extract migrated under reducing conditions. (figure 5, compare the two left western blots). This demonstrates that XRCC3 underwent modification upon UVA irradiation, which prevents antibodies to recognize the protein when migrating in non-reducing conditions. Since modified XRCC3 is detected by the antibodies when migrating under reducing conditions, it means that two or more cysteins were oxidized upon UVA irradiation and that it is reversible by the reducing agent β-mercaptoethanol. We also found that singlet oxygen produced by UVA radiation causes this reversible disulfide bound in XRCC3. Furthermore, we observed that this phenomenon is quickly reversed in cells.

While the consequences of irreversible (PCNA) or reversible (XRCC3) oxidation on the functionality of these proteins has not yet been fully investigated, our data emphasize that protein oxidation may play a role in the deleterious effect of UVA radiation.

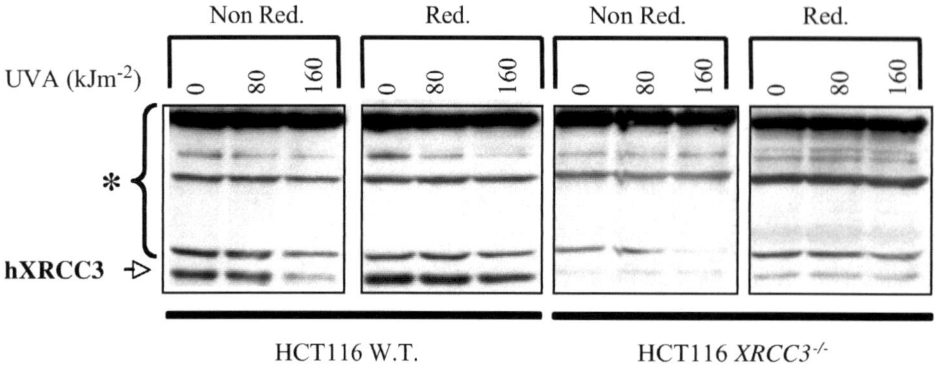

Figure 5. XRCC3-proficent and deficient human HCT116 cells were exposed to increasing doses of UVA radiation. Immediately post radiation, cells were lysed on ice in lysis buffer containing 4 mM N-ethylmaleimide (NEM) an irreversible thiol alkylating agent that protects cysteine residues from oxidation during lysis. Equal amounts (30 mg) of soluble protein extracts in non reductive (without β-mercaptoethanol) and reductive (150 mM β-mercaptoethanol) conditions were loaded on 9 % SDS-PAGE and transferred onto nitrocellulose membrane. Human Xrcc3 protein was detected using rabbit polyclonal anti-Xrcc3 antibody (Novus Biological, clone 100-165). The star indicates non specific cross-reactivity of the antibody.

5. Conclusion

Excessive exposure to sunlight is a major cause of skin cancer but also contributes to premature skin aging. The two main components of sunlight that contribute to photocarcinogenesis and photoaging are UVB and UVA radiation, respectively. While it is known for long time that UVB is a major source of DNA damage, more recent data highlight also a key role of UVA radiation in the formation of CPDs, 8-oxoguanine and strand breaks. Efficient repair of such DNA damage is crucial for the maintenance of genomic stability. In other words, inability of cells to efficiently carry out repair could lead to mutations and consequently to genomic instability, a marker of cancer cells. Interestingly, UVA-induced ROS can also oxidize DNA repair proteins (i.e. PCNA, XRCC3, Oggl) raising the possibility that during sunlight exposure, there is a narrow window during which DNA repair proteins could be oxidatively modified, and possibly inactivated, and DNA damages are produced. For a better understanding of the deleterious effects of UVA, it seems therefore essential, not only to measure the various type of DNA lesions induced by UVA (or simulated sunlight) but also to measure the extent of protein oxidation with particular emphasis on DNA repair enzymes.

Acknowledgements

We are grateful to Dr. Aziz Sancar for photolyase and Dr. R Stephen Lloyd for T4 endonuclease V. This work was supported by Centre National de la Recherche Scientifique, Institut Curie, a grant from the French "Agence Nationale pour la Recherche" (ANR-07-PCVI-0004-01) to E S. and by a grant from the National Cancer Institute of Canada (NCIC) (with funds from the Canadian Cancer Society and the Terry Fox Run) to R D. D G. is the recipient of a doctoral fellowship from "La région Ile de France" on DIM SenT program. R D. holds the Canada Research Chair in "Genetics, Mutagenesis and Cancer".

References

[1] Sutherland J C and K P Griffin 1981 *Rad. Res.* **86** 399.
[2] Tyrrell R M and Pidoux M 1989 *Photochem.Photobiol.* **49** 407.
[3] Basu-Modak S and Tyrrell R M 1993 *Cancer Res.* **53** 4505.
[4] Ryter S and Tyrrell R M 1998 *Free Radical Biol. Med* **24** 1520.
[5] El Ghissassi F, Baan R, Straif K, Grosse Y, Secretan B, Bouvard V, Benbrahim-Tallaa L, Guha N, Freeman C, Galichet L and Cogliano V 2009 *Lancet Oncol.* **10** 751.
[6] Ridley A J, Whiteside J R, McMillan T J and Allison S L 2009 *Int. J. Rad. Biol.* **85** 177.
[7] Tyrrell R M 1973 *Photochem. Photobiol.* **17** 69.
[8] Tyrrell R M, Webb R B 1973 *Mutat Research* **19** 361.
[9] Tyrrell R M, Ley R D and Webb R B 1974 *Photochem. Photobiol.* **20** 395.
[10] Peak M J and Peak J G 1990 *Photochem. Photobiol.* **51** 649.
[11] Zhang X S, Rosenstein B S, Wang Y, Lebwohl M, Mitchell D M and Wei H C 1997 *Photochem.Photobiol.* **65** 119.
[12] Kvam E and Tyrrell R M 1997 *Carcinogenesis* **18** 2379.
[13] Douki T, Perdiz D, Grof P, Kulucsics Z, Moustacchi E, Cadet J and Sage E 1999 *Photochem.Photobiol.* **70** 184.
[14] Besaratinia A, Kim S and Pfeifer G P 2008 *FASEB J.* **22** 2379.
[15] Kielbassa C, Roza L and Epe B 1997 *Carcinogenesis* **18** 811.
[16] Pouget J P, Douki T, Richard M J and Cadet J 2000 *Chem.Res. Toxic.* **13** 541.
[17] Douki T, Reynaud-Angelin A, Cadet J and Sage E 2003 *Biochem.* **42** 9221.
[18] Young A R, Potten C S, Nikaido O, Parsons P G, Boenders J, Ramsden J M and Chadmick C A 1998 *J. Inves. Derm.* **111** 936.
[19] Perdiz D, Grof P, Mezzina M, Nikaido O, Moustacchi E and Sage E 2000 *J.Biol. Chem.* **275** 26732.
[20] Rochette P J, Therrien J P, Drouin R, Perdiz D, Bastien N, Drobetsky E A and Sage E 2003 *Nucl. Acids Res.* **31** 2786.
[21] Courdavault S, Baudouin C, Charveron M, Favier A, Cadet J and Douki T 2004 *Mutation Res.* **556** 135.
[22] Mouret S, Baudouin C, Charveron M, Favier A, Cadet J and Douki T 2006 *Proc. Nat. Acad. Sci. USA* **103** 13765.
[23] Cadet J, Sage E and Douki T 2005 *Mutation Res.* **571** 3.
[24] Girard P M, Pozzebon M, Delacôte F, Douki T, Smirnova V and Sage E 2008 *DNA Repair* **7** 1500.
[25] Wischermann K, Popp S, Moshir S, Scharfetter-Kochanek K, Wlaschek M, de Gruijl F, Hartschuh W, Greinert R, Volkmer B, Faust A, Rapp A, Schmezer P and Boukamp P 2008 *Oncogene* **27** 4269.
[26] Lamola A A 1968 *Photochem.Photobiol.* **7** 619.
[27] Charlier M and Helene C 1972 *Photochem. Photobiol.* **15** 527.
[28] Kuluncsics Z, Perdiz D, Brulay E, Muel B and Sage E 1999 *J. Photochem. Photobiol B: Biol.* **49** 71.
[29] Jiang Y, Rabbi M, Kim M, Ke C, Lee W, Clark R L, Mieczkowski P A and Marszalek P E 2009 *Biophys. J.* **96** 1151.
[30] Schuch A P, da Silva Galhardo R, de Lima-Bessa K M, Schuch N J and Menck C F 2009 *Photochem. Photobiol. Sci.* **8** 111.
[31] Mouret S, Philippe C, Gracia-Chantegrel J, Banyasz A, Karpati S, Markovitsi D and Douki T 2010 *Org. Biomol. Chem.* **8** 1706.
[32] Bensasson R V, Land E J and Truscott T G 1993 Oxford University Press, Oxford (UK)
[33] Davies M J 2003 *Biochem.Biophys. Res. Comm.* **305** 761.
[34] Pattison D I and Davies M J 2006 L.P. Bignold editor Birkhaüser verlag (Switzeland) 131-157
[35] Shacter E 2000 *Methods Enzymol.* **319** 428.

[36] Tyrrell R M and Sage E 2011 *Handbook of Photochemistry and Photobiology* 3rd edition CRC Press in the press
[37] Tyrrell R M, Webb R B and Brown M S 1973 *Photochem. Photobiol.* **18** 249.
[38] Vile G F and Tyrrell R M 1995 *Free Rad. Biol. Med.* **18** 721.
[39] Maresca V, Flori E, Briganti S, Camera E, Cario-André M, Taïeb A, and Picardo M *J. Invest. Dermat* **126** 182.
[40] Ogawa F, Sander C S, Hansel A, Oehrl W, Kasperczyk H, Elsner P, Shimizu K, Heinemann S H and Thiele J J 2006 *J. Invest. Dermat.* **126** 1128.
[41] Bravard A, Vacher M, Gouget B, Coutant A, de Boisferon F H, Marsin S, Chevillard S and Radicella J P 2006 *Mol. Cell. Biol.* **26** 7430.
[42] Eiberger W, Volkmer B, Amouroux R, Dhérin C, Radicella J P and Epe B 2008 *DNA Repair* **7** 912.
[43] Dardalhon D, Angelin A R, Baldacci G, Sage E and Francesconi S 2008 *Cell Cycle* **7** 611.
[44] Montaner B, O'Donovan P, Reelfs O, Perrett C M, Zhang X, Xu Y Z, Ren X, Macpherson P, Frith D and Karran P 2007 *EMBO Rep.* **8** 1074.

COST Chemistry CM0603–MELUSYN Joint Meeting IOP Publishing
Journal of Physics: Conference Series **261** (2011) 012003 doi:10.1088/1742-6596/261/1/012003

Oxidation of protein tyrosine or methionine residues:
From the amino acid to the peptide

J Bergès[1,2,3], P Trouillas[4] and C Houée-Levin[5,6]
[1]Université Pierre et Marie Curie, UMR 7616, Laboratoire de Chimie Théorique, 75005 Paris, France - [2]CNRS, UMR 7616, Laboratoire de Chimie Théorique, 75005 Paris, France - [3]Université Paris Descartes, 75006 Paris, France -[4]EA 4021 Faculté de Pharmacie, 2 Rue du Dr. Marcland, 87025 Limoges Cedex, France - [5]Université Paris Sud, UMR 8000, Laboratoire de Chimie Physique, 91405 Orsay (France). [6]CNRS 91405 Orsay, France

E-mails: jb@lct.jussieu.fr - patrick.trouillas@unilim.fr - chantal.houee@u-psud.fr

Abstract. Methionine and tyrosine are competing targets of oxidizing free radicals in peptides or proteins. The first step is the addition of OH radicals either on the sulphur atom of methionine, followed by OH^- elimination, or on the aromatic cycle of tyrosine. The next step can be stabilization of methionine radical cation by a two centre-three electron bond, or intramolecular electron transfer from tyrosine to the methionine radical cation. In this latter case a tyrosine radical is formed, which appears deprotonated. In a first step we have compared the stability of the OH radical adducts on Methionine or on Tyrosine. In agreement with experimental results, the thermodynamical data indicate that the OH adduct on Tyrosine and the radical cation are more stable than those on methionine. In a second step we have investigated the stabilization of the radical cations of Methionine by formation of intramolecular S∴X two-center three-electron bond (X=S, N, O). Finally we have compared the spin densities on separated amino acids to that in a radical pentapeptide, methionine enkephalin. One observes a delocalisation of the orbital of the odd electron on the sulfur atom of Met and on the cycle of Tyr. The peptidic chain is also concerned.

1. Introduction

In oxidative stress, among other species the very strong oxidant OH radicals are formed. Their high reactivity renders them extremely deleterious. We are currently investigating the one-electron oxidation of peptides and proteins by experimental and theoretical methods. Experiments often show that among the most important targets of these radicals are the Methionine (Met) and Tyrosine (Tyr) residues. In particular, in neurodegenerative involved peptides and proteins (β amyloid in Alzheimer's disease, α -synuclein in Parkinson's disease, Prion protein), the oxidation of Methionine to its sulfoxide is a key event in their conformational change and thus in the development of the diseases. On the other hand, oxidation of Tyr residues in the natural opioid peptides enkephalins, leads to an increase of pain sensation.

The main lines of the mechanisms of the oxidation of these two residues are fairly well known (figure 1). In Methionine the first step would be the addition of OH radical followed by OH^- elimination leading to the sulfur centred radical cation. The radical cation could be then stabilized by

Journal of Physics: Conference Series **261** (2011) 012003 doi:10.1088/1742-6596/261/1/012003

forming a two-centre three-electron (2c-3e) bond with any electron -donating atom [1]. As for tyrosine, the first step is also the addition of OH on the cycle, but OH^- elimination is slow at neutral pH. With other oxidants it can undergo oxidation giving its radical cation that deprotonates. When Methionine and Tyrosine are simultaneously present in a peptide, oxidation of methionine may be followed by intramolecular reduction by a tyrosine residue ending by the regeneration of Met and the formation of a tyrosinyl deprotonated radical. . Depending on the way that is followed, the final compounds are peptide dimer or peptide with DOPA or with oxidized methionine.

Using various radicals as oxidants, this process was characterized in methionine enkephalin [2] (Figure 2), model peptides [3] and ribonuclease A [4]. Very recently this process was reinvestigated in model dipeptides [5] and it was proposed that intramolecular electron transfer needed some proximity of the residues. Preliminary results suggested that a tyrosyl residue (Tyr-125) adjacent to a methionine residue (Met-127) (figure 1) in alpha-synuclein underwent electron transfer. Conversely in thioredoxin [6], calmodulin [7], amyloid peptide [8], the methionyl 2c-3e radical was observed and no long -range intramolecular electron transfer was noticed. In another model system, it was suggested that methionine could serve as a stepping stone between an electron acceptor and Tyr as a donor [9,10]. Clearly, the role of methionine on protein tyrosyl oxidation deserves further investigation.

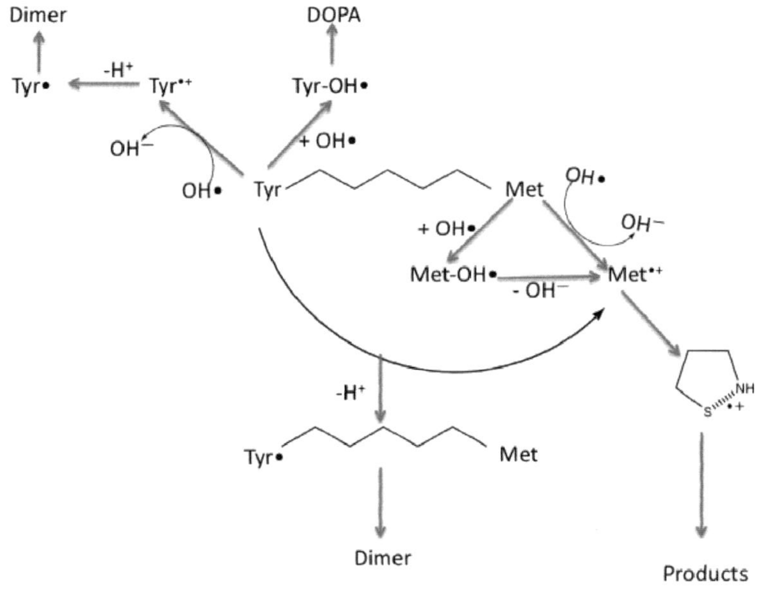

Figure 1. Main steps of the oxidation by OH radicals of a peptide containing Met and Tyr.

In this paper, our aim is to make a preliminary thermodynamical analysis of the oxidation of a typical peptide containing Methionine and Tyrosine, Methionine enkephalin (Met enk) (figure 3) by OH radicals. In this peptide, the intramolecular step was not rate-limiting because it was too fast [11], and is noticed only by the appearance of the tyrosinyl deprotonated radical. We chose among the 80 structures in the Protein Data Bank (code 1PLW) a geometry in which Methionine and Tyrosine are close to each other (figure 2). This geometry is similar to that encountered in peptides involved in neurodegenerative diseases (e.g. in alpha synuclein). We have already studied the OH addition to both amino acids [12], thus in this paper we considered the amino acids as residues, i.e. we blocked both the carboxylate and the amine functions (figure 3).

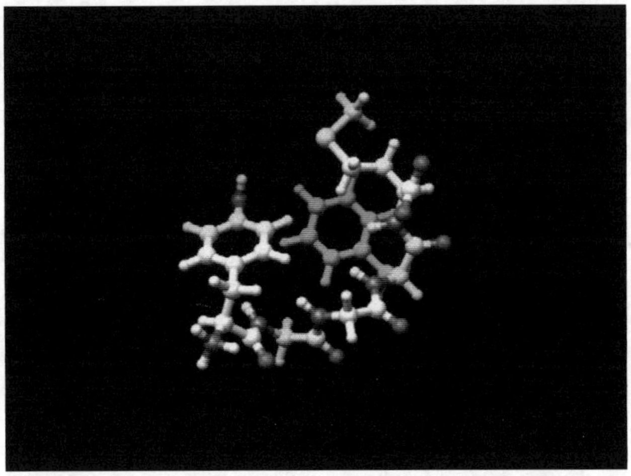

Figure 2. Methionine enkephalin (structure 50, PDB code 1PLW).

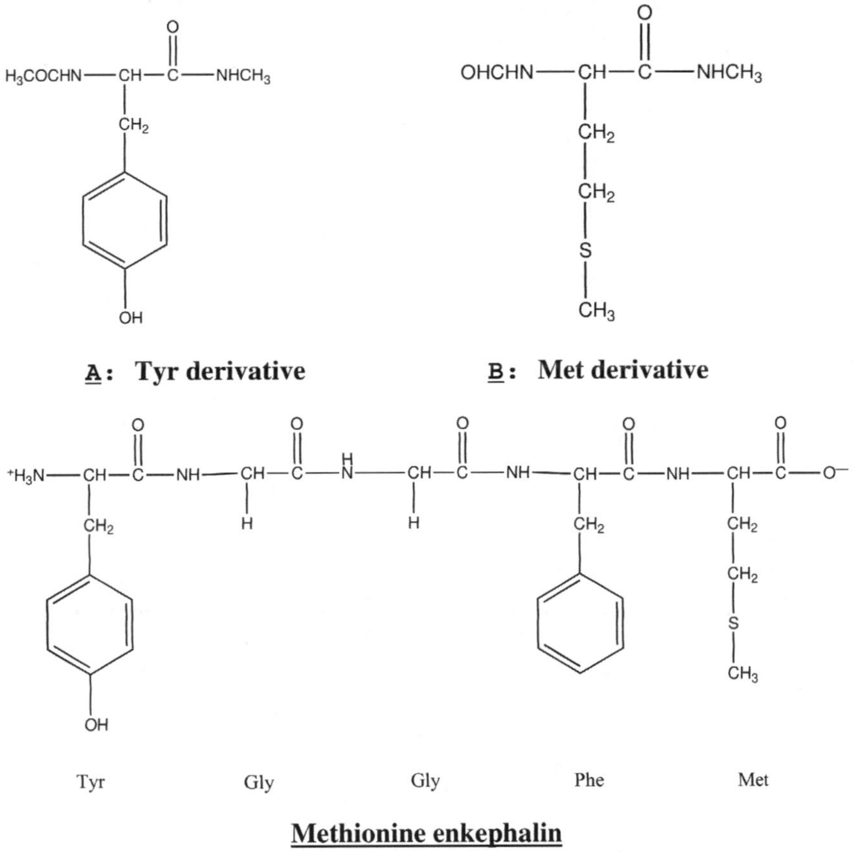

Figure 3. The compounds considered in this study.

We began the study by considering the first step of the oxidation, i.e. the addition of OH on Met and Tyr separately, and the direct electron transfer reaction between OH radicals and the amino acids (reactions (1) and (2)).

$$\text{Tyr (or Met)} + \text{OH} \rightarrow \text{Tyr-OH (or Met-OH)} \quad (1)$$

$$\text{Tyr (or Met)} + \text{OH} \rightarrow \text{Tyr}^{\bullet+} \text{ (or Met}^{\bullet+}\text{)} + \text{OH}^{-} \quad (2)$$

Then we examined some possible fates of Met radical cations. (i) The formation of intramolecular 2c -3e bond that stabilized the entity as it was suggested several times. This was done in some dipeptides. (ii) The fate of the methionine radical if such a stabilization does not occur because of geometrical constraints. Thus we analysed the various conformations of Met enkephalin and selected one in which the sulfur atom is not close to any nitrogen or oxygen atom that could possibly make a 2c-3e bond. This conformation was optimised and the ionisation and the oxidation reactions were studied like for the separated amino acids.

2. Computational methods

The geometries, enthalpies (ΔH) and Gibbs energies (ΔG) of the derivatives of Tyr **A** and Met **B** (Figure 3) and products (their adducts and radical cations) were calculated with (U)B3P86/6-31+G(d,p). Enthalpies and Gibbs energies took into account temperature-dependent corrections (zero point energy (ZPE), translational, rotational and vibrational energies) at 298K. All the different ground states were confirmed by vibrational frequency analysis, i.e. no imaginary frequency. The same method was used for the Met Enkephaline and its radical cation.

As for the 2c-3e radical cations the bond lengths and the binding energies are overestimated except with the BH&HLYP functional [13]. We thus privileged the BH&HLYP functional in the spin-unrestricted formalism and used the standard 6-31G(d) basis set. The use of this relatively small basis set is justified by the fact that DFT methods are not very basis-set dependent. All structures were fully optimized using the analytical gradient technique, and the nature of each located stationary point was checked by evaluating harmonic frequencies. When considered, the effect of solvation was taken into account using a continuum model (CPCM). The drawings of Figure 2 and 8 were obtained using the free software Chimera [14].

3. Results

1.1. OH adducts on Tyr and Met

ΔH (and ΔG) of the addition reaction was $\Delta H^{addition} = \Delta H\text{ (adduct)} - \Delta H\text{(molecule)} - \Delta H(^{\bullet}OH)$. The $^{\bullet}$OH addition was thermodynamically favoured on the aromatic ring of Tyr. Depending on the substituent the reaction Gibbs enthalpy ranged from -16 to -10 kcal/mol in vacuum and in water. The most stable stereoisomer of Tyr-OH adduct is shown in figure 4.

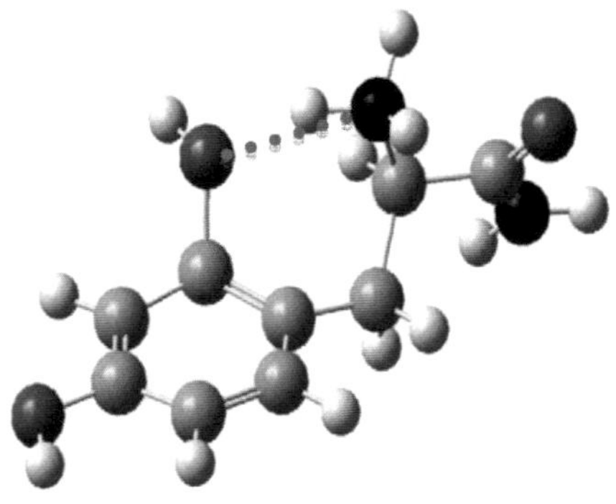

Figure 4. The most stable configuration of the OH adduct on Tyrosine is stabilized by an intramolecular hydrogen bond (dotted line) between the OH group and the amine function.

As for methionine, the OH-adduct is less favoured (Gibbs energy of -3.7 kcal/mol in vacuum), which is in agreement with experimental results by pulse radiolysis (i.e. very low (microsecond) and long (millisecond) lifetimes respectively for Met-OH and Tyr-OH adducts respectively [15]). This trend is confirmed in water.

The spin densities are mostly on the cycle in the Tyr adduct and shared between S and O of OH atoms in Methionine (figure 5). The delocalization in Tyr adducts explains the greater stability of these adducts.

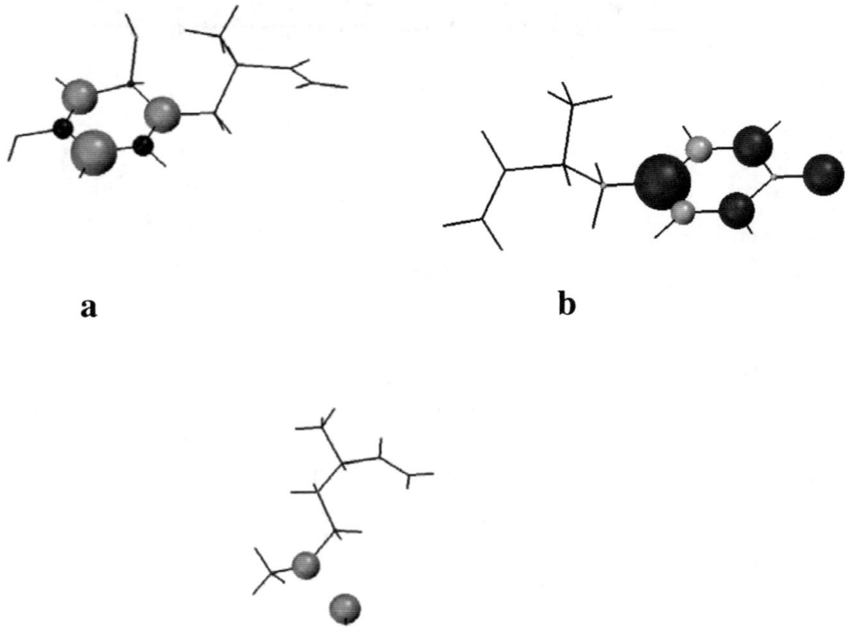

a **b** **c**

Figure 5. Spin densities on the OH adduct (a), on the tyrosinyl radical (b), on methionine OH adduct (c).

1.2. Radical cations of Tyr and Met derivatives

Both radical cations are stable structures. The one formed from **B** is shown in figure 6. The ionisation potentials (IP) of both molecules are gathered in table 1.

Table 1. Ionisation potentials of the compounds **A**, **B** and **Met enk**. (see figure 3).

Compound	**A**	**B**	**Met enk**
Ionisation Potential (eV)	8.40	8.75	7.53

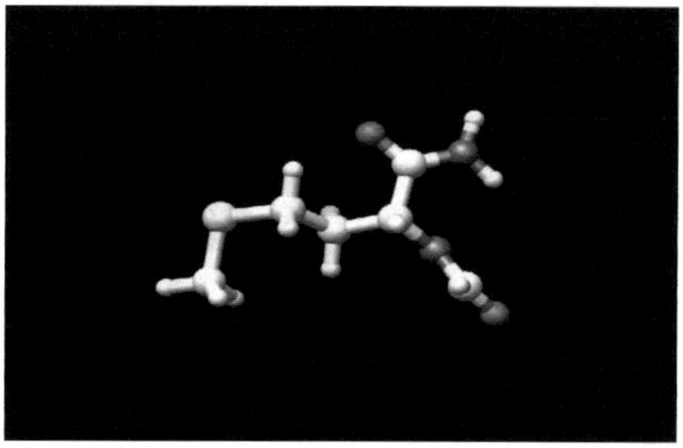

Figure 6. Radical cation from methionine derivative **B**.

Both IP are of the same order of magnitude, as expected, however the one of **A** is lower than that of **B**. The spin densities are splitted on several atoms. In **B** the spin density is mostly on the sulfur atom (0.62). In **A**, the spin density is mostly on the cycle (0.70, on the -COH bond and the C in para position).

As for the reaction of oxidation of each residue by OH radicals (reaction (2)) the energy variations are 141.9 and 150.1 kcal/mol respectively for **A** and **B** in vacuum. They follow the order of ionisation potentials. Including solvation strongly decreases these values as expected but the order remains the same. Globally it appears that it the reaction of OH is easier with the Tyr derivative than with the Met one, for the electron transfer as well as for the addition reaction. In Tyrosine, experimental results indicate that H atom abstraction from the OH group of Tyr by OH radical is not favoured. Conversely the radical cation undergoes very fast deprotonation of the phenol giving the tyrosinyl radical. With other oxidants, H atom transfer either direct or indirect (i.e., via (i) coupled electron-proton transfers or (ii) adduct formation followed by H_2O release) leads to the same tyrosinyl radical. The O-H bond dissociation enthalpy (BDE) of Tyr is around 90 kcal/mol. The radical is stabilized by the spin density delocalization over the entire aromatic ring (figure 5). The spin density on the O-atom from where the H-atom was removed is relatively low (around 0.37).

1.3. Stabilization of the Met radical cations in peptides

The radical cations of Met could be stabilized by 2c 3e bonds with N or O atoms of the vicinal peptidic bond. The most stable are 5 membered cycles involving nitrogen of the peptidic bond or 6 membered rings of the peptidic bond (figure 7) [16]. The SN bonded entity is more stable than the SO one by 16 kcal/mol. It means that even with this stabilisation, the oxidation of Tyr is more favourable energetically.

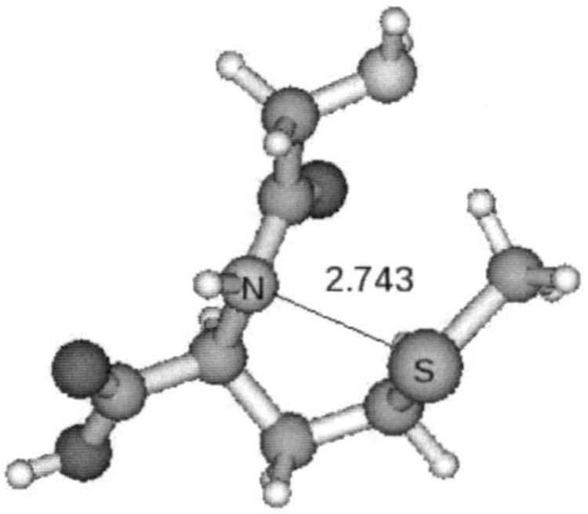

Figure 7. Stabilisation of the sulfur- centred radical cation by a 2c-3e bond making a pseudo cycle of 5 atoms.

1.4. Insight into the one-electron oxidation of Met enkephalin

There are 80 experimental conformations obtained by NMR in micelles (PDB code 1PLW). A conformational analysis of the enkephalins was performed [17,18]. The peptide backbone exists as a mixture of folded and unfolded forms (approximately 50% each). In the folded ones the Tyr and Met residues are close to each other (distance S...OH around 3.5 Å). We extracted some typical conformations that we have optimized and we show here the results obtained with a folded one (structure 50, figure 2), distance S...OH = 3.3 Å). The radical cation was optimised (figure 8). In this cation the folding has increased, but the distance S...O(phenol) remained of the same order of magnitude (3.8Å). This relatively large distance indicates that no 2c-3e bond was created between these two atoms.

The ionisation potential is much lower than that of the separated amino acids. Similarly the energy variation of the reaction of oxidation corresponding to Reaction (2) is lower than for the amino acids (121kcal/mol, Table 1). The spin density is remarkably delocalised. The S atom bears only 16% of the total spin density, 15 % is found on the cycle of Tyrosine and 30 % on the N-terminal amine, the rest being dispersed in the molecule.

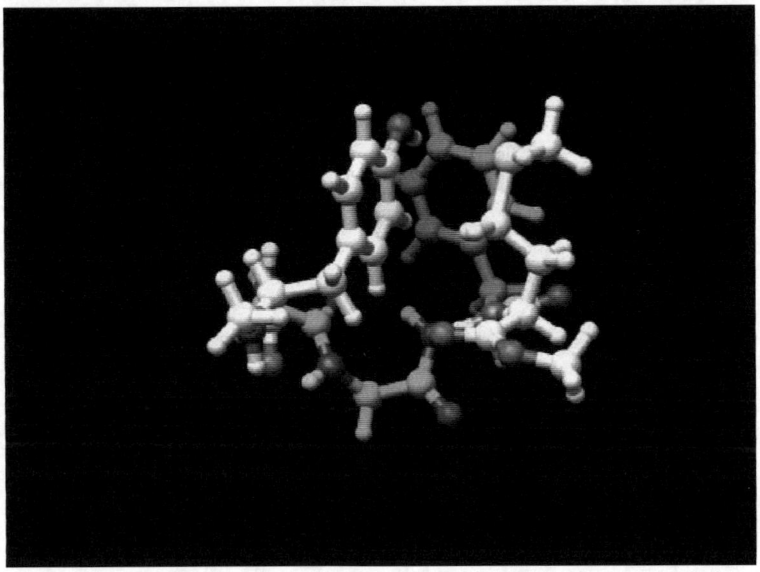

Figure 8. Optimised structure of Methionine enkephalin radical cation (starting point: structure 50 of PDB 1PLW, see figure 2).

4. Conclusion

In this work we compared the one-electron oxidation of two protein residues, tyrosine and methionine by OH radicals in derivatives and in a pentapeptide, Methionine enkephalin. In general, from a thermodynamical point of view, ˙OH radicals may add more favourably on the aromatic rings of Tyr than on Met in agreement with results by pulse radiolysis. The radical cations are stable structures. However their spin densities are somewhat delocalised. Especially in the pentapeptide that we studied, the major part of the spin density is neither on the sulfur nor on the tyrosine residue. More investigations are necessary to understand the structure of the radicals coming from one electron abstraction on peptides.

Acknowledgments

We thank the COST CM0603 action for fruitful discussions.

Journal of Physics: Conference Series **261** (2011) 012003 doi:10.1088/1742-6596/261/1/012003

References

[1] Clark T. 1990 in *Sulfur-centered reactive intermediates in Chemistry and Biology*, Chatgilialoglu C.; Asmus D., Eds.; NATO-ASI Studies, Plenum Press: New York; vol. 197 13.

[2] Bobrowski K,. Wierzchowski K L, Holcman J and Ciurak M 1992 *Int J Radiat Biol.* **62** 507.

[3] Bobrowski K,. Wierzchowski K L, Holcman J and Ciurak M 1990 *Int J Radiat Biol.* **57** 919.

[4] Prütz W A, Butler J, Land E J 1985 *Int J Radiat Biol Relat Stud Phys Chem Med.* **47** 149.

[5] Zhang H, Zielonka J, Sikora A, Joseph J, Xu Y and Kalyanaraman B 2009 *Arch. Biochem. Biophys.* **484** 134.

[6] Sicard-Roselli C, Lemaire S,. Jacquot J-P, Favaudon V, Marchand C and. Houée-Levin C 2004 *Eur. J. Biochem.* **271** 3481.

[7] Nauser T, Jacoby M, Koppenol W H, Squier T C and Schöneich C 2005 *Chem. Commun.* **5** 587.

[8] Kadlcik V, Sicard-Roselli C, Mattioli T A, Kodicek M. and Houée-Levin C 2004 *Free Radical Biol. Med.* **37** 881.

[9] Giese B, Wang M, Gao J, Stoltz M, Muller P and Graber M 2009 *J. Org. Chem.* **74** 3621.

[10] Wang M, Gao J, Muller P and Giese B 2009 *Angew. Chem. Int. Ed.* **48** 4232.

[11] Mozziconnacci O 2008 *Ph. D. thesis, Université Paris Sud.*

[12] Trouillas P, Bergès J, Houée-Lévin C. 2010 *Int. J. Quantum Chem.* DOI 10.1002/qua.22556.

[13] Braïda B, Hiberty P, Savin A **1998**. *J. Phys. Chem. A* **102** 7872.

[14] Pettersen E F, Goddard T D, Huang C C, Couch G S, Greenblatt D M, Meng E C, Ferrin T E 2004 *J Comput Chem.* **25** 1605.

[15] Asmus K. D **2001**. in *Radiation Chemistry, present status and future trends* Jonah C D and. Rao B S M eds. Elsevier, Amsterdam, 341.

[16] Fourré I, Bergès J and Houée-Levin C 2010 *J. Phys. Chem. A* **114** 7359.

[17] Vengadesan K and. Gautham N 2004 *Biopolymers* **74** 476.

[18] Hansmann U H.E, Okamoto Y and Onuchic J N 1999 *Proteins* **34** 472.

Photosensitization of plasmid-DNA loaded with platinum nano-particles and irradiated by low energy X-rays

E Porcel[1], K Kobayashi[3], N Usami[3], H Remita[2], C Le Sech[1] and S Lacombe[1]
[1.] Université Paris-Sud 11, CNRS, Institut des Sciences Moléculaires d'Orsay-ISMO (UMR 8214), 91405, Orsay Cedex, France, [2.] Université Paris-Sud 11, CNRS, Laboratoire de Chimie Physique (UMR 8000), 91405 Orsay Cedex, France, [3.] Photon Factory, Institute of Materials Structure Science, High Energy Accelerator Research Organization, Oho, Tsukuba, Ibaraki 305-0801, Japan.

Email: claude.lesech@u-psud.fr

Abstract. Damage in DNA plasmids (pBR322) loaded with platinum nanoparticles (NP-Pt) DNA-NP and irradiated with monochromatic X-rays tuned to the resonant photoabsorption energy of the L_{III} and M_{III} electronic inner-shell of platinum - respectively 11556 eV and 2649 eV - and off-resonant X-rays - 11536 eV and 2639 eV- is investigated. In all the experiments, an enhancement of the single and double strand break - SSB and DSB - yields is observed when NP-Pt are present. Amplification effects are almost similar for the irradiations performed at on and off the L or M shell resonance suggesting that a non resonant mechanism is responsible for the major part of the DNA breaks enhancement.The amount of DNA breaks measured in the present work is compared to the results in similar experiments made with complexes of plasmid-DNA containing platinum molecule : chloroterpyridine platinum (PtTC). The average number of PtTC molecules in the solution is the same as in the experiments made with NP-Pt in order to study a possible difference in the radiosensitization efficiency when the high-Z atoms are clustered (NP-Pt) or dispersed in the system (PtTC). A mechanism is suggested involving photoelectrons which can efficiently ionize the platinum atoms. These results are consistent with those observed when DNA-NP complexes are irradiated by fast atomic ions. These findings suggest that any nanoparticle made of high-Z atoms might behaves as radiation enhancer whatever the ionizing radiation is electromagnetic or charged particle source.

1. Introduction

The Auger effect is widely used in radiobiology as a tool to generate secondary electrons and charged centers in order to induce molecular damages in the surrounding biomolecules. The underlying idea is that these damages will induce cell death and as a consequence might improves the therapeutic index in radiotherapy. Electron emitting radionucleides such as ^{125}I, ^{131}I, ^{32}P are currently used in the treatment of thyroid tumors or leukemia. Many papers are devoted to the subject and they cannot all be listed here. We refer to a recent review article and references therein [1].

When an inner-shell excitation is induced in an atom, the relaxation of the electronic excited core can occur by photons emission or electron emission - Auger effect -. Both processes exist at the same time, and the branching ratio is in favour of Auger effect for K-shell excitation of relatively low Z atoms ($Z < 30$) or L, M, N … electronic shell excitation for larger values of the atomic number. For

Journal of Physics: Conference Series **261** (2011) 012004 doi:10.1088/1742-6596/261/1/012004

example the energy relaxation of the L_{III} shell of platinum (Z = 78) is about 80% via the Auger effect, whereas it is only 3% following the relaxation of a K shell excitation in this atom.

Different possibilities exist to induce a hole in an atomic inner shell. The capture of a K shell electron by the nucleus in excess of protons is the mechanism responsible for the Auger effect in the radionucleide ^{125}I. A more common way to trigger the Auger effect is to induce the excitation of an electron belonging to a K, L, M ... shell by photoelectric effect or electronic impact. The absorption of photons by atoms depends of their energies and may exhibit a resonance structure near the ionization threshold. A particular atomic species in a molecule can thus be selected as a target for the primary energy deposition by tuning the photon energy to a specific resonant process.

In the present work we present the results on DNA breaks induction by low energy photons tuned to the L_{III} (11556 eV) and M_{III} (2649 eV) shell of platinum when plasmid DNA are loaded with high-Z atoms included in a molecule (PtTC) or with nanoparticles of platinum (NP-Pt). In the off-resonant X-rays experiments, the energy of the photons was adjusted below the resonant absorption peaks corresponding to the L_{III} and M_{III} shell respectively at 11536 eV and 2639 eV. In addition some experiments were made in presence of a free radicals scavenger, the dimethylsulfoxide (DMSO), in order to disentangle the respective contribution of the indirect effects - mediated by free radicals attacks - from the non mediated by free radicals 'direct effects'.

2. Materials and methods

2.1. Preparation of the samples

For a simple and rapid quantification of simple and complex damage, plasmid DNA pBR322 provided by TOYOBO (Japan) is used in our study. The double-stranded DNA of 4361 base pairs (2.83×10^6 Daltons) is diluted in TE buffer (10 mmol L^{-1} Tris-HCl (pH = 7.6) and 1 mmol L^{-1} ethylenediaminetetraacetic acid (EDTA)). Plasmid DNA presents the advantage to have three conformations namely supercoiled, circular, and linear with respectively no break, single strand break, double strand break, which can be separated by migration in agarose gel electrophoresis. Prior to irradiation, DNA samples contain more than 95 % supercoiled, 5 % circular and no linear forms.

Two types of radiosensitizers are studied in the present work. The first one was the platinum terpyridine chloride (PtTC), used as a platinum containing molecule, whose effect has been studied for long by the group [2]. The molecule PtTC is not an intercalator and does not induce any artefact in the analysis of the plasmids at the concentrations used in this work. It interacts smoothly with DNA through electrostatic interaction and is located close to the DNA. It is thus a good candidate to probe the physical processes, induced at the atomic level in the high-Z atoms bound to the DNA, triggered by X-rays irradiation.

The PtTC molecules were obtained from FLUKA - Sigma Aldrich Chemie Gmbh Schnelldorf Germany - and used without any further purification. As shown previously by the group, a ratio of 1 platinum atom per 15 phosphate atoms (7 or 8 base pairs) corresponds to the optimized quantity of platinum to get efficient sensitization [2]. The concentrations of PtTC solutions were determined by spectrophotometry, at λ = 278 nm. The absorption coefficient ε is equal to 25100 $L.mol^{-1}.cm^{-1}$ [3] at this wavelength.

Platinum nanoparticles (NP-Pt) coated with polyacrylic acid (PAA Sigma-Aldrich) were synthesized by radiolytic reduction of platinum complexes $Pt(NH_3)_4Cl_2H_2O$ (Sigma Aldrich10^{-4} mol L^{-1}) in aqueous solution containing PAA 0.5 mol L^{-1}. For more details in the protocol see reference [4]. The solutions were deaerated prior to the irradiation by bubbling with nitrogen. The irradiation was carried out in a panoramic source ^{60}Co source, at a dose rate of 2.2 kGy h^{-1} (1 Gray = 1 J kg^{-1}). The irradiated solutions were protected from light and stored at 4°C. Transmission electron microscopy (TEM) observations were performed on a JEOL JEM 100 CXII transmission electron microscope at accelerating voltage of 100 kV. The nanoparticules have almost the same size, about 3 nm, and each contains around 1000 atoms. The solutions were finally diluted down to a concentration in metallic atoms similar to the concentration reached with PtTC.

A solution of DNA and PtTC or DNA and nanoparticles was prepared with a defined ratio r (ratio r = number of Pt atoms/number of P atoms in DNA). In the present work we have taken r = 1/15. No study with different ''r'' values have been made in the present work. Such study can be found in a previous work [5] and shows that the number of DSB increases along the values of r. The DNA was incubated with radiosensitizer for one hour prior to irradiation. When PtTC molecules or nanoparticles NP-Pt were added to the samples the concentrations of platinum atoms remain constant. The concentration of DNA and salty species remains also constant for every samples. The addition of any of the radiosensitizers had no deleterious effects on DNA, as shown systematically with controls. In order to investigate the role of the water radicals, the free radical scavenger DMSO was added in some samples at a concentration equal to 1.0 mol.L^{-1}.

2.2. Irradiations
Irradiations by X-Rays were performed at the Photon Factory on beamlines 27 A and B, Tsukuba, Japan. A sample chamber for irradiation was made of acrylic resin with a Kapton film (7.9 μm) window. The window area was 8 mm (in height) x 10 mm (in width) and the depth was 2 mm. A sample solution of 100 micro liter was put into the vessel. After appropriate doses of irradiation under an aerobic condition, small aliquots were taken out of the sample, and analyzed with gel electrophoresis. More details about the experimental setup can be found in the references [2, 6]. The doses ranged from 0 to 250 Grays when DMSO is not present and up to 800 Gy in presence of DMSO. Dose rates are a few Gy/min at BL-27 A and about 10 Gy/min at BL-27 B.

2.3. Plasmids analysis
The samples were loaded with 1 μL loading dye solution. The electrophoresis was performed in a 1.7 % agarose gel, in an electric field of 10 V cm^{-1}, at 4°C during 3h30. After migration, the gel was stained with ethidium bromide (1μg mL^{-1}), and the DNA lines were revealed under UV light (302 nm) and recorded by a CCD camera. The free of access image analysis software (National Institute of Health Image) was used to quantify the intensity of DNA lines of the different DNA conformations. Considering that the supercoiled plasmid (S) is binding the ethidium bromide 1.42 times less than the relaxed (R) and the linear (L) forms, the percentage of the three different conformations of DNA is determined. The average fraction of single and double strand breaks per plasmid (SSB, DSB) is calculated according to an earlier report [7]. In the equations below L and S represent respectively the normalized amount of linear and supercoiled forms of plasmids.

$$SSB = \ln[(1 - L)/S] \text{ and } DSB = L/(1 - L) \quad (1)$$

3. Results and discussion
It is important to remember that the average number of platinum atoms is the same in all experiments presented here. The numbers of SSB and DSB per plasmid as a function of the dose (figures 1-4) increases in a linear way. The numerical values of the slope of the lines resulting from the fit are reported in the table 1. In all experiments an enhancement of the number of SSB and DSB is observed when the plasmids are in presence of NP-Pt or PtTC.

The figure 1 displays the induction of SSB in free DNA and complexes of DNA loaded with PtTC molecules or NP-Pt particles following irradiation by photons tuned to the M_{III} energy shell of platinum. The law of DNA breaks induction can be considered to be linear. The number of breaks is almost the same whether the photon energy is tuned to the M_{III} shell threshold (2649 eV) or just below (2639 eV). The number of SSB induced in similar conditions in DNA complexes with PtTC molecules is slightly larger than the latter one. The results obtained in experiments with PtTC irradiated with photons at 2639 eV (off resonance) are not shown for sake of clarity. They have also almost the same values than those measured at 2649 eV.

Journal of Physics: Conference Series **261** (2011) 012004 doi:10.1088/1742-6596/261/1/012004

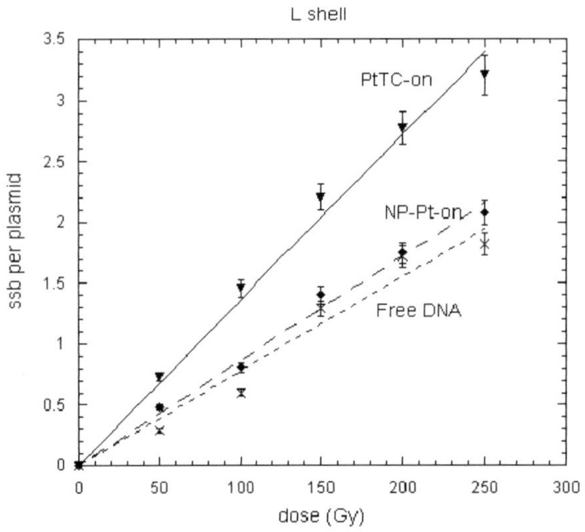

Figure 1. Number of SSB per plasmid versus dose induced by photons tuned on and off the M_{III} inner-shell energy. Full triangles stand for irradiation of plasmid DNA added with PtTC. Open squares and diamonds stand for irradiation of plasmids added with nano-particles. Crosses represent the number of single breaks in free DNA.

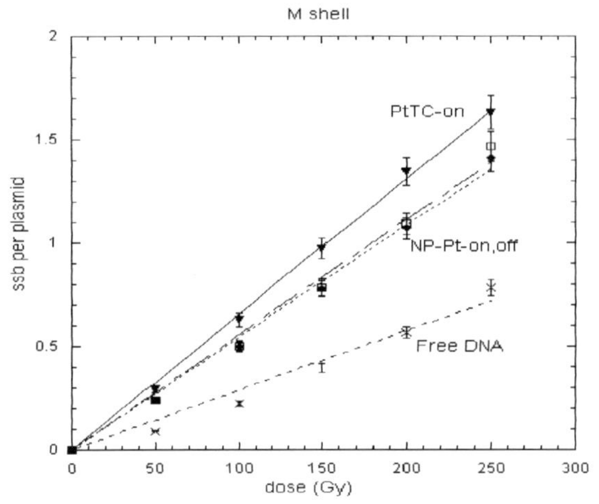

Figure 2. Number of SSB versus dose induced in plasmid by photons tuned on the L_{III} inner-shell energy. Full triangles stand for irradiation of plasmid DNA added with PtTC. Diamonds stand for irradiation of plasmids added with nano-particles. Crosses represent the number of single breaks in free DNA.

The figure 2 above displays the results on the SSB induction by photons at an energy corresponding to the L_{III} energy shell of platinum. We can check in the table 1 that in all experiments the number of SSB per plasmid at the energy corresponding to the L_{III} shell is larger than the number of SSB induced by photons tuned to the M_{III} shell energy. However, when the nanoparticles are present, the specific enhancement Δm (equation 2) is smaller at this energy than to the value for the M_{III} shell energy. We propose below an explanation of this finding.

$$\Delta m = m_{SSB}(DNA + nanos) - m_{SSB}\,(free\ DNA) \qquad (2)$$

Figures 3 and 4 show the number of DSB induced by photons tuned respectively to the M_{III} and L_{III} shell energy. The law for the DSB induction is seen to be linear and the numerical values of the slopes are reported in the table 1. It is important to notice that for all experiments made in presence of platinum atoms contained in PtTC molecules or nanoparticles the number of DSB is increased significantly.

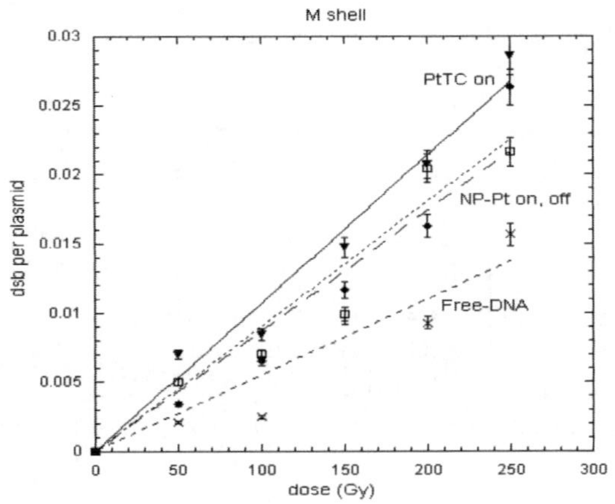

Figure 3. Representation of the DSB versus dose induced by photons tuned to the M_{III} inner-shell energy. Full triangles stand for irradiation of plasmid DNA added with PtTC. Open squares and diamonds stand for irradiation of plasmids added with nano-particles. Crosses represent the number of double breaks in free DNA.

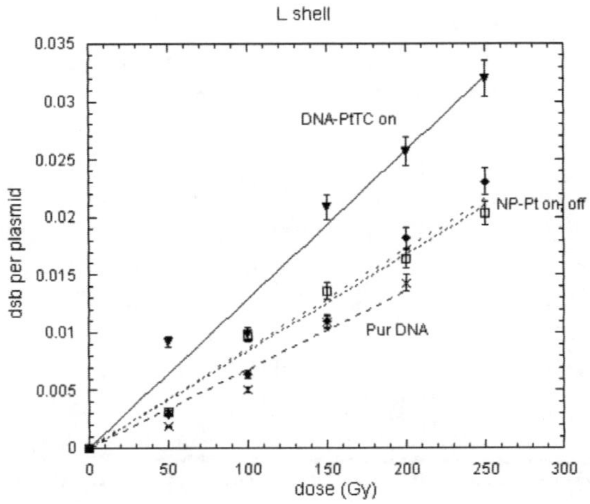

Figure 4. Representation of the DSB versus dose induced by photons tuned to the L_{III} shell energy. Full triangles stand for irradiation of plasmid DNA added with PtTC. Open squares and diamonds stand for irradiation of plasmids added with nano-particles. Crosses represent the number of double breaks in free DNA.

The values reported in table 1 show that the number of DSB induced at L_{III} or M_{III} shell energy are close to each other and have almost the same values when the DNA is added with nanoparticles ($9x10^{-5}$ breaks per plasmid per Gray). The values of the DSB-L shell obtained in previous experiments [2, 8] have been also reported in table 1. The values of reference 2 are relative to previous experiments with photons at L_{III} energy and similar to the present one and show a good agreement with the present measures. The induction of DSB with C^{6+} carbon ions is studied in [8]. In the latter experiments with C^{6+} ions the values of the slope in presence of PtTC ($19x10^{-5}$) and without ($9x10^{-5}$) are larger than the slope measured when photon are used ($13x10^{-5}$, $6.7x10^{-5}$). However one can remark that in all cases - DSB-M shell, DSB-L shell and DSB-C^{6+} - the ratio m(DSB, PtTC)/m(free DNA) is always near 2, independent of the photon or particle energy.

Table 1. Numerical values of the slopes m of the lines in the figures 1 – 4

	PtTC on	Nanos on	Nanos off	Free DNA
SSB-L shell x10^{-3}	13.6 $\pm$ 3.4x10^{-1}	8.6 $\pm$ 1.9x10^{-1}	8.4 $\pm$ 1.9x10^{-1}	7.7 $\pm$ 3.7x10^{-1}
SSB-M shell x10^{-3}	6.5 $\pm$7x10^{-2}	5.5 $\pm$ 10^{-1}	5.4 $\pm$ 10^{-1}	2.9 $\pm$ 10^{-1}
DSB-L shell x10^{-5}	13 $\pm$ 5x10^{-1}	9 $\pm$ 4x10^{-1}	8.4 $\pm$ 2x10^{-1}	6.7 $\pm$ 4x10^{-1}
DSB-L shell [2]	12.7x10^{-5}			7.4x10^{-5}
DSB-M shell x10^{-5}	10.6 $\pm$ 4x10^{-1}	9 $\pm$ 6x10^{-1}	8.6 $\pm$ 6x10^{-1}	5.5 $\pm$ 5x10^{-1}
DSB C^{6+} [8]	19x10^{-5}			9x10^{-5}
SSB-M shell x10^{-4} DMSO	6.8 $\pm$ 5x10^{-1}	6.4 $\pm$ 4x10^{-1}	7.4 $\pm$ 5x10^{-1}	5.7 $\pm$ 4x10^{-1}
DSB-M shell x10^{-5} DMSO	1.4 $\pm$ 10^{-1}	1.5$\pm$ 1.2x10^{-1}	1.5 $\pm$ 1.2x10^{-1}	1.6 $\pm$ 1.2x10^{-1}

In the table 1 we report also the slopes of the SSB and DSB induction in presence of DMSO at the concentration of 1 M. The large decrease of the values of the slopes indicates that the SSB and DSB are mainly due to the attack of free radicals hydroxyl radicals $HO^{\bullet}$ following radiolysis of water induced by secondary electrons. The Auger effects in platinum atoms induced by secondary electrons, like photoelectrons, results mainly in $HO^{\bullet}$ free radicals production [2].

It is generally admitted that the SSB are induced by free radicals $HO^{\bullet}$ that have migrated through the medium, whereas DSB are the consequence of production of dense free radicals in the proximity of the DNA [9]. The stronger Auger effect - more energy relaxed in L_{III} shell than in the M_{III} shell - induces more free radicals $HO^{\bullet}$. The local recombination of the $HO^{\bullet}$ radicals in a dense cluster tends to decrease the number of migrating $HO^{\bullet}$, with the consequence that less SSB breaks are induced. Such discussion has been already presented in reference [10].

The Auger effect in L_{III} or M_{III} shell occurring in proximity of the DNA plasmid is always sufficient to induce a DSB. It can be said that there is some kind of saturation of DSB induction in that case. The number of DSB induction due to nanoparticles is less than the induction due to PtTC molecules. It should not be concluded that PtTC molecules are more efficient than nanoparticles to generate double DNA breaks. In the present experiments the nanoparticles have not the same binding capacity to DNA in comparison to the PtTC molecules as mentionned above. They are statistically located at a larger distance than PtTC from DNA. This different location in the medium is less favorable to induce the DSB as already mentioned [9].This simple model explains well why the specific enhancement of SSB due to nanoparticles decreases when the energy of the photons is larger, whereas the number of DSB is practically independent of the photon energy, at least in the present range of energies. The proposed model interprets also the fact that the major part of the SSB is scavengeable by DMSO due to the migration of $HO^{\bullet}$ in the buffer, whereas the DSB are less scavengeable because the $HO^{\bullet}$ produced close to DNA are less easily scavenged by DMSO.

The nanoparticles are widely considered nowadays to be a possible approach to improve the therapeutic index of radiotherapy. The present experiments suggest that no large difference in DNA damages induction is observed when nanoparticles (NP-Pt) or PtTC molecules are used. The soft X-rays used here will hardly present any interest for medical protocols in radiotherapy because they are strongly absorbed by the matter and cannot penetrate into the tissues. Radiation using X-rays in the 100-200 keV have been proposed by different groups see for example [11] and references therein. These photons can penetrate in a deeper location in tissues. However they are strongly absorbed in the entrance channel and can induce severe damages like radiodermitis in the healthy tissues.

In our group we have shown that the combination of radiation with ions (He^{2+}, C^{6+}) and PtTC or NP-Pt can enhance the cell death when they are loaded with PtTC [8] or DSB when plasmids [12] are added with NP-Pt. The atomic ions have an interesting dose deposition profile, preserving the entrance channel and they are able to deliver a large dose at a deep seated location (Bragg peak). For this reason they should be particularly efficient in the combination with radiosensitizers such as nanoparticles made with high-Z atoms. The key point is to design nanoparticles of high-Z atoms (platinum, gold …) able to target with a good selectivity the tumor. A great deal of works toward this goal is made in different groups. For example in the reference [13] one can find different considerations about this challenge.

4. Conclusion

Most of the experiments dealing with the study of radisensitization by nanoparticles made of heavy atoms (platinum, gold) make use of photons X-rays in the 100 keV energy range. The present experiments demonstrate that low energy electron can also trigger the enhancement of DNA breaks by a non resonant process mediated by secondary electrons like photoelectrons. The mechanisms involving secondary electrons presented here suggests that any ionising radiation can induce the activation of heavy atoms contained in nanoparticles or molecules via the secondary electrons emitted along the track of the ionising particles. This hypothesis is confirmed by experiments made in similar conditions with fast carbon ions C^{6+} particles (290MeV/amu).

Considering the property of dose deposition profile in matter by the atomic ions, the analysis presented in this work suggests that the combination of irradiation by ions and suitable nanoparticles presenting a significant selective uptake in tumor might be an efficient tool to improve the therapeutic index of radiotherapy.

References

[1] Kobayashi K, Usami N, Porcel E, Lacombe S and Le Sech C 2010 *Mutation Res. Rev.* **704** 123.
[2] Kobayashi K, Frolich H, Usami N, Takakura K and Le Sech C 2002 *Rad. Res.* **157** 32.
[3] Jennette K W, Lippard S J, Vassiliades A, and Bauer W R 1974 *Proc. Nat. Acad. Sci* USA**71** 3839.
[4] Belloni J, Mostafavi M, Remita H, Marignier J L and Delcourt M O 1998 *New J. Chem.* **22,** 1239.
[5] Le Sech C, Takakura K, Saint-Marc C, Frohlich H, Charlier M, Usami N, and Kobayashi K 2001 *Can. J. Physiol. Pharm.* **79** 196.
[6] Maeda M, Kobayashi K, Hieda K 2004 *IJRB* **80** 841.
[7] Spotheim-Maurizot M, Charlier M and Sabattier R 1990 *Int. J. Rad. Biol.* **57** 301.
[8] Usami N, FurusawaY, Kobayashi K, Lacombe S, Reynaud-Angelin A, Sage E, Wu T D, Croisy A, Guerquin-Kern J L and Le Sech C 2008 *Int. J. Rad. Biol.* **84** 603.
[9] Kassis AI, Harapanhalli R S and Adelstein S J 1999 *Rad. Res.* **151** 167.
[10] Usami N, Hirayama R, Kobayashi K, Furusawa Y, Porcel E, Lacombe S and Le Sech C 2010 *J. Rad. Res.* **51** 21.
[11] Hébert E M, Debouttière P J, Lepage M, Sanche L and Hunting D J 2010 *Int. J. Rad. Biol.* **86** 692.
[12] Porcel E, Liehn S Remita H, Usami N, Kobayashi K, Furusawa Y, Le Sech C and Lacombe S 2010 *Nanotechnology* **21** 085103.
[13] Chithrani D B, Jelveh S, Jalali F, van Prooijen M, Allen C, Bristow R G, Hill R P and Jaffray D A 2010 *Rad. Res.* **173** 719.

Bacterial and archaeal resistance to ionizing radiation

F Confalonieri and S Sommer
Univ. Paris-Sud, CNRS UMR8621, Institut de Génétique et Microbiologie, Bâtiments 400-409, Université Paris-Sud, 91405 Orsay, France

E-mail: fabrice.confalonieri@u-psud.fr - suzanne.sommer@u-psud.fr

Abstract. Organisms living in extreme environments must cope with large fluctuations of temperature, high levels of radiation and/or desiccation, conditions that can induce DNA damage ranging from base modifications to DNA double-strand breaks. The bacterium *Deinococcus radiodurans* is known for its resistance to extremely high doses of ionizing radiation and for its ability to reconstruct a functional genome from hundreds of radiation-induced chromosomal fragments. Recently, extreme ionizing radiation resistance was also generated by directed evolution of an apparently radiation-sensitive bacterial species, *Escherichia coli*. Radioresistant organisms are not only found among the Eubacteria but also among the Archaea that represent the third kingdom of life. They present a set of particular features that differentiate them from the Eubacteria and eukaryotes. Moreover, Archaea are often isolated from extreme environments where they live under severe conditions of temperature, pressure, pH, salts or toxic compounds that are lethal for the large majority of living organisms. Thus, Archaea offer the opportunity to understand how cells are able to cope with such harsh conditions. Among them, the halophilic archaeon *Halobacterium sp* and several *Pyrococcus* or *Thermococcus* species, such as *Thermococcus gammatolerans*, were also shown to display high level of radiation resistance. The dispersion, in the phylogenetic tree, of radioresistant prokaryotes suggests that they have independently acquired radioresistance. Different strategies were selected during evolution including several mechanisms of radiation byproduct detoxification and subtle cellular metabolism modifications to help cells recover from radiation-induced injuries, protection of proteins against oxidation, an efficient DNA repair tool box, an original pathway of DNA double-strand break repair, a condensed nucleoid that may prevent the dispersion of the DNA fragments and specific radiation-induced proteins involved in radioresistance. Here, we compare mechanisms and discuss hypotheses suggested to contribute to radioresistance in several Archaea and Eubacteria.

1. Introduction

Extreme environments are a permanent challenge for living organisms. However, a large diversity of microorganisms belonging to the three domains of life (bacteria, archaea, eukaryotes) were isolated from harsh environments such as hydrothermal vents, surface sands of hot arid deserts exposed to intense ultraviolet (UV) radiation, cycles of extreme temperatures, and desiccation [1]. Artificial environments, such as radiation and toxic chemical waste dumps, also provide intense selection pressure for extremophiles.

Among these highly stress resistant organisms, bacteria belonging to the *Deinococcaceae* family are known for their ability to survive exposure to ionizing radiation and *Deinococcus radiodurans* is, by far, the best characterized. It was first isolated in canned meat exposed to 4,000 Gy γ-irradiation in order to achieve sterility [2]. This bacteria is characterized by an exceptional ability to withstand the lethal effects of DNA-damaging agents, including ionizing radiation, ultraviolet light and desiccation [3]. *D. radiodurans* can survive extremely high doses of radiation such as 5,000 Gy γ-irradiation generating approximately 200 DNA double-strand breaks, 3000 DNA single-strand breaks and more than 1000 damaged bases per genome, without loss of viability. Survival is also not affected by a dose of 500 J / m^2 UV-light that generates up to 5,000 pyrimidine dimers in DNA, and *D. radiodurans* shows 85% viability after 2 years in the presence of less than 5% humidity. *D. radiodurans* is also able to deal with high oxidative stress. However, resistance to ionizing radiation is not a common trait of all *Deinococcus* species. Indeed, some *Deinococcales* species, such as *Deinococcus radiomollis*, *Deinococcus claudionis*, *Deinococcus altitudinis* and *Deinococcus alpinitundrae*, isolated recently from alpine environments, were shown to be sensitive to ionizing radiation [4]. Moreover, a survey of bacteria identified eleven phyla that contain species with unusually high resistance to the lethal effects of ionizing radiation. These phyla are not closely related to each other and do not share a recent common lineage [5].

Radioresistance is also widespread among hyperthermophilic archaea. Among them, *Thermococcus gammatolerans* is almost as radioresistant as *D. radiodurans* (figure 1). This anaerobic organism was isolated after a γ-ray exposure at a dose of 30 kGy, of an enriched culture of microorganisms collected in a North Pacific hydrothermal vent [6]. When cells grow in a rich medium, they can withstand a dose of 5000 Gy without lethality [7] (figure 1). This property is not surprising since at high temperature, which corresponds to the natural environment of this archaeon, DNA accumulates many types of damage (e.g., hydrolytic depurination, deamination of cytosine and adenine, and single- or double-strand breaks) that have to be very rapidly repaired to avoid complete denaturation of DNA. However, a 5,000 Gy dose of ionizing radiation drastically decreases the viability of *Pyrococcus abyssi*, a hyperthermophilic archaeon phylogenetically close to *T. gammatolerans* with optimal growth temperatures approaching the boiling point of water (figure 1). Consequently, the radioresistance of hyperthermophilic organisms cannot be explained only through their adaptation to high temperatures. Despite the differences observed in the *Thermococcales* order, *Pyrococcales* species, compared to most living species, can be considered as radioresistant, showing 100 % viability at a dose of 2,000-2,500 Gy.

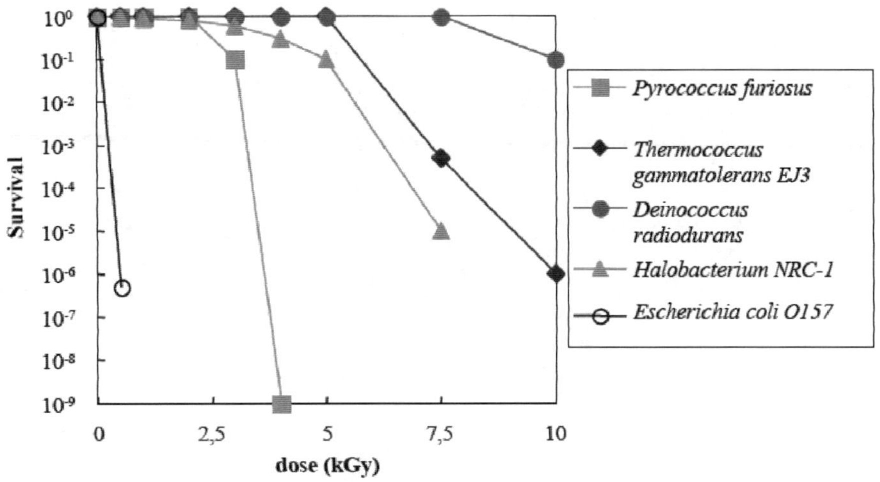

Figure 1. Survival curves for *Pyrococcus furiosus*, *T. gammatolerans*, *D. radiodurans*, *Halobacterium* and *E. coli*, following exposure to γ irradiation.

In contrast to *Thermococcales*, the halophilic archaeon *Halobacterium* sp. NRC-1 is aerobic, mesophilic and obligatory halophilic, requiring a salt concentration 10 times higher than those found in sea-water (250 g/L versus 20 g/L) for growth. A high intracellular KCl concentration together with a membrane pigmented by bacterioruberin provide protection from DNA damaging agents [8]. It has a capacity to cope with desiccation, UV radiation and ionizing radiation, its cell survival being 80% after irradiation at a dose of 2500 Gy and 10% at 5000 Gy [9].

More recently, Harris *et al* [10] have isolated 5 independent derivatives of the model *Escherichia coli* bacteria almost as radioresistant as *D. radiodurans* by exposure of the radiosensitive bacteria to multiple cycles of increasing radiation doses and selection of the survivors after each cycle of irradiation. The existence of so many unrelated radioresistant species suggests that the molecular mechanisms that protect against ionizing radiation-induced damage evolved independently in these organisms and that they combine a variety of physiological tools that are tightly coordinated to achieve radioresistance. In this review, we will discuss and compare different features that have been proposed to contribute to radioresistance in these Bacteria and Archaea.

2. Effects of ionizing radiation on living cells

Ionizing radiation, produced by the decay of radioactive elements, leads to the formation of multiple ions and electrons. γ-rays are photons modifying many molecules in a cell and producing highly reactive hydroxyl radicals (ROS), in particular $OH^{\cdot}$ by ionization of water molecules. These free radicals cause, directly or indirectly, a variety of DNA damage products, such as base modifications, DNA single-strand and double-strand breaks. It is generally admitted that 80% of the DNA damage result indirectly by the action of ROS and only 20% from direct interaction between γ photons and DNA [11]. Bacteria or Archaea highly resistant to ionizing radiation, are not protected against DNA damage [12] and the number of DNA double strand breaks per cell at a given dose is proportional to the size of the genome. However, we did not notice a correlation between resistance to radiation and the size of the genome, the number of chromosomes or the number of genes encoded by these organisms (table 1 and figure 1).

Table 1. Genomic information in different microorganisms [13-17] tested for their sensitivity to ionizing radiation (see figure 1)

Species	*Escherichia coli*	*Pyrococcus furiosus*	*Halobacterium NRC-1*	*Thermococcus gammatolerans*	*Deinococcus radiodurans*
Domain	Bacteria	Archaea	Archaea	Archaea	Bacteria
Genome size (Megabases)	4.64	1.91	2.57	2,04	3.28
Chromosomes and plasmids	1 chromosome	1 chromosome	1 chromosome and 2 megaplasmids	1 chromosome	2 chromosomes, 1 megaplasmid and 1 plasmid
Number of open reading frames	4289	2065	2630	2157	3187
% GC	50,8	40,7	65.9	54,0	66,6

In addition to causing DNA injuries, elevated doses of ionizing radiation also provoke damage to many other cell components.

3. Protection against oxidative stress

The first line of protection against oxidative stress is the presence of antioxidant enzymes such as superoxide dismutases and catalases that protect biomolecules from ROS-mediated damage (figure 2A). Superoxide dismutase catalyzes the conversion of oxygen superoxide to hydrogen peroxide which is subsequently transformed into H_2O by either catalases or peroxidases (figure 2A). The latter enzymes are coupled to redox pathways such as the ascorbate/glutathione cycles found in plants [18]. *D. radiodurans*, encodes 3 predicted superoxide dismutases and 3 predicted catalases [19] induced after exposure to ionizing radiation.

In Archaea, superoxide dismutase is mostly encoded by aerobic organisms (like *Halobacterium*) and few anaerobic organisms such as *Methanosarcina barkeri* [20]. In other anaerobes, this enzyme is replaced by a superoxide reductase (SOR) (figure 2B) [21-23]. Superoxide reductase oxidized during the processing of ROS needs to be reduced by rubredoxin which is in turn re-oxidized by NAD(P)H rubredoxin oxidoreductase (NROR). Finally hydrogen peroxide is most likely converted into H_2O by the action of the rubrerythrin and possibly by as-yet uncharacterized peroxidases.

In addition, some anaerobic organisms can use another pathway to eliminate superoxides without production of hydrogen peroxide (figure 2C) [24]. A crystallographic study revealed that superoxide reductase forms a complex with ferrocyanides [25]. This complex reacts very efficiently with oxygen radicals, the end product of the reaction being not hydrogen peroxide, but oxidized formate and H_2O (figure 2C).

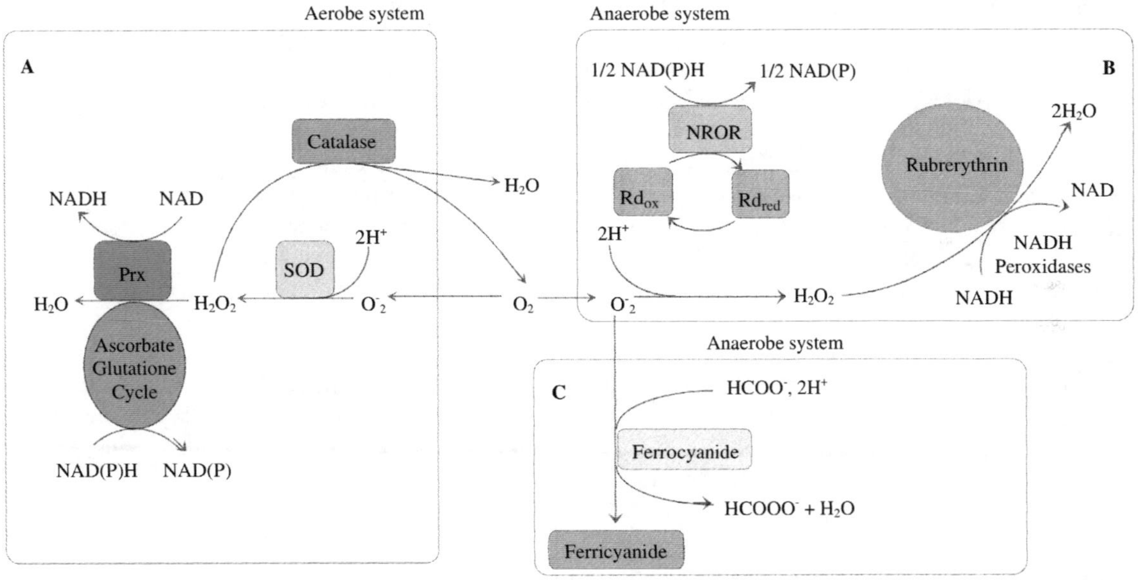

Figure 2. Proposed pathways for detoxification of oxygen superoxide in : (A) aerobic and (B and C) anaerobic organisms. Cyt bd: cytochrome bd ubiquinol oxidase. Prx: peroxiredoxin, Rd: rubredoxin, SOD: superoxide dismutase, SOR: superoxide reductase, NROR: NAD(P)H rubredoxin oxidoreductase.

A second line of protection against oxidative stress is the presence of an enhanced [Mn]/[Fe] ratio that confers an increased capacity to prevent the formation of iron-dependent ROS through the Fenton reaction ($H_2O_2 + Fe^{2+} \rightarrow 2\ OH^{\cdot} + Fe^{3+}$) [26]. Intracellular Mn can also be protective by scavenging ROS [27]. The most radioresistant species contain an intracellular concentration of Mn about 300 times higher and an intracellular concentration of Fe about 3 times lower than the most-sensitive species [28]. When *D. radiodurans* cells were grown in defined medium without Mn supplementation, a 1000-fold reduction in survival was observed at 10,000 Gy γ-irradiation. Mn(II) does not directly prevent DNA double-strand breaks [29], but it can mimic the activities of catalase and superoxide dismutase [30].

An inverse correlation between the [Mn]/[Fe] concentrations ratios and the levels of protein oxidation was shown, with no protein oxidation being detected in cells with the highest [Mn]/[Fe] ratios [28]. It was also reported that desiccation resistant dry-climate soil bacteria accumulate high intracellular manganese and low iron concentrations compared to desiccation sensitive bacteria. Moreover, their proteins are protected from oxidation during drying [31].

It is interesting to note that a high intracellular [Mn]/[Fe] ratio was also found in the radioresistant archaeon *Halobacterium NRC-1* [32]. *P. furiosus* increased the expression of a ferritin/Dps-like gene after irradiation, and the protein by chelating iron, could limit the production of hydroxyl radicals by the Fenton reaction [33]. Therefore, the decrease of iron levels in a cell could be a common strategy between some radioresistant bacteria and archaea to counteract protein oxidization. However, this feature cannot be generalized since the [Mn]/[Fe] ratio in several *E. coli* radioresistant mutants is the same than those in wild type cells [10].

Other ions also play a protective role against oxidative stress. A high concentration of NaBr (1.7 M) enhances cell survival of *Halobacterium* NRC-1 to ionizing radiation [32]. This improved ability to respond to ionizing radiation is due to bromine ions, which traps hydroxyl radicals, thereby significantly reducing the oxidation of nucleotides and proteins when cells are irradiated.

However, whereas the levels of protein oxidation are not the same in the radioresistant *D. radiodurans* bacteria and the radiosensitive *Escherichia coli* bacteria, the number of DNA double strand breaks generated in their DNA is the same even if 80% of the DNA damage result indirectly by the action of ROS [12]. This led Daly and collaborators to postulate that protection of proteins could be a major feature of radioresistance, as cells require functional proteins present immediately after irradiation for DNA repair [28]. Interestingly, efficient protection against protein carbonylation produced by cellular metabolism, or induced by radiation, was also observed in *E. coli* strains generated by directed evolution of *E. coli* K-12 and specifically adapted to survive exposure to high doses of ionizing radiation [34].

It was also recently proposed that proteins with abundant hydrophilic low complexity (LC) regions participate to the recovery processes of *D. radiodurans* to DNA damage and to its resistance to desiccation [35].

4. Efficient DNA double-strand break repair

4.1. Kinetics of DNA double-strand break repair

DNA double-strand breaks are considered the most dangerous and difficult to repair DNA lesions. However, *D. radiodurans* bacteria and *T. gammatolerans* archaea can sustain γ-irradiation doses that introduce hundreds of DNA double-strand breaks in their genome. The kinetics of *D. radiodurans* DNA double-strand break repair after exposure to 6,800 Gy γ-irradiation is very rapid as an intact genome complement is reconstructed from a myriad of fragments in two to three hours when cells where incubated post-irradiation in a rich-nutrient medium (figure 3).

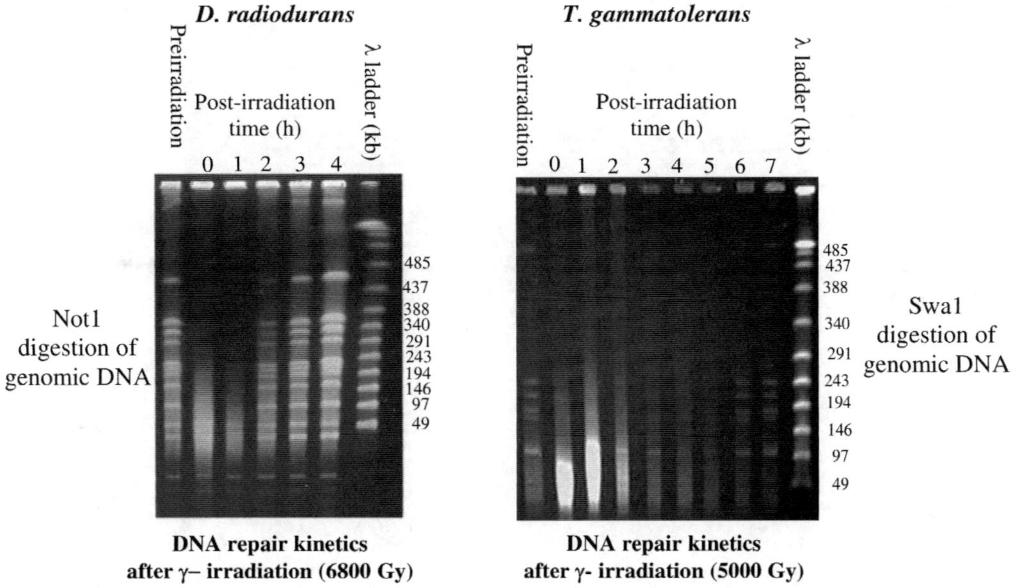

Figure 3. Kinetics of genome reconstitution in *D. radiodurans* and *T. gammatolerans* cells exposed to γ-ray. *D. radiodurans* and *T. gammatolerans* exponentially growing cells were exposed to 6,800 Gy or 5000 Gy γ-irradiation, respectively, and then incubated in a rich culture medium at 30°C for *Deinococcus* or at 85°C for *Thermococcus* with agitation for recovery. To avoid double-strand breaks generated by mechanical manipulation of the DNA and not present *in vivo*, cells were immobilized and lysed in agarose plugs at different post-irradiation times. The plugs were then incubated with proteinase K, an enzyme that destroys proteins, and RNAse to eliminate the RNA. Finally, the purified DNA was digested using a rare cutting restriction enzyme (Not1 or SwaI) before being analyzed by pulsed field gel electrophoresis. Detailed protocols are available in [7,36].

4.2. Mechanisms involved in DNA double-strand break repair

Several mechanisms have been proposed to account for such an efficient DNA double-strand break repair (figure 4). To repair DNA double strand breaks by homologous recombination (HR), single strand annealing (SSA), or extended synthesis dependent strand annealing (ESDSA), cells need to contain another intact copy of the damaged DNA region. In contrast, non-homologous end joining (NHEJ) does not require a second homologous copy of DNA to join two contiguous fragments [37].

4.2.1. Homologous recombination (HR). HR is the main pathway involved in DNA double-strand break repair in bacteria [39] and in the yeast *Saccharomyces cerevisiae* [40]. HR uses an intact homologous DNA molecule to restore the correct DNA sequence at damaged sites. The first step of DNA double-strand break repair by homologous recombination is the processing of the DNA ends to generate single stranded 3' overhangs on which the recombinase RecA in bacteria, RadA in Archaea [41] and Rad51 in eukaryotic cells are recruited to form a nucleoprotein filament. The next steps of HR are invasion of a homologous overlapping fragment, exchange of DNA strands, extension of the nascent DNA heteroduplex and resolution of the resulting crossover structure.

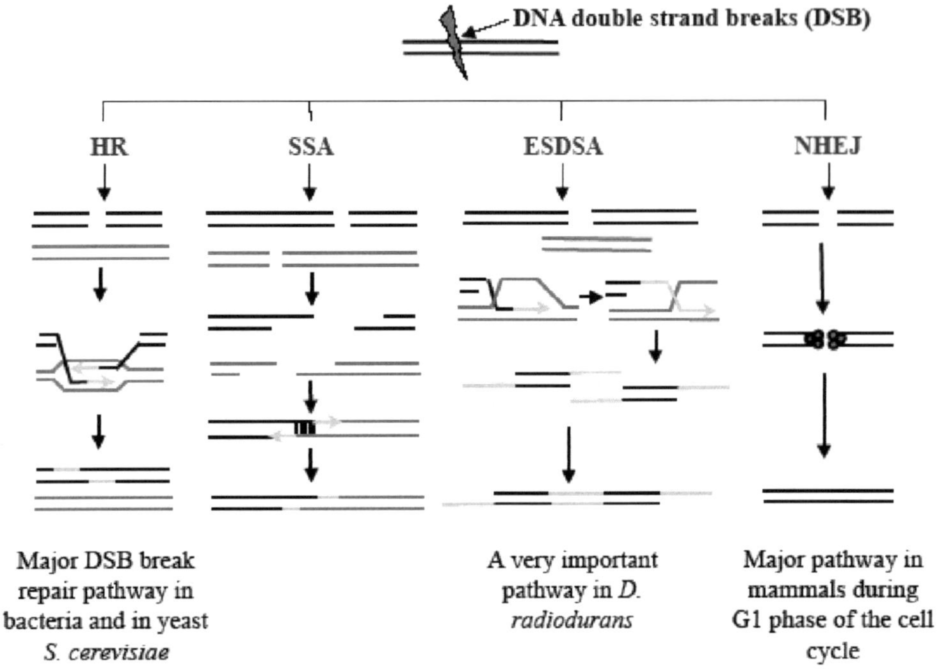

Figure 4. Different pathways for DNA double-strand break repair (adapted from [38]). Green lines indicate DNA newly synthesized during DNA double-strand break repair. Red and black are used to represent two copies of the same chromosomal region.

4.2.2. Single-Strand Annealing (SSA). In addition to homologous recombination, SSA has been proposed to occur at early times in irradiated Deinococcal cells to account for the observation that part of the radiation-induced double-strand breaks can be mended in a recombination-defective *recA* mutant [42]. The process of SSA can occur if two DNA double-strand breaks are located in the same region of two copies of the same chromosome. After exonuclease-catalysed resection of DNA ends, single-stranded overhangs are produced. If the overhangs contain complementary sequences, they can anneal. Then, single stranded regions present in the reconstituted chromosome are filled in by DNA synthesis (figure 4).

4.2.3. Extended Synthesis-Dependent Strand Annealing (ESDSA). Recently, Zharadka et al (2006) [43] proposed a variant of the SSA model called ESDSA to account for their findings that (i) DNA fragments assembly coincides with a massive DNA synthesis which occurs at a much higher rate in irradiated cells than in unirradiated growing cultures (ii) high amount of single-stranded DNA are transiently present in the damaged cells (iii) the reassembled genome appears to be composed of a patchwork of contiguous blocks of old (in black on figure 4) and newly synthesized DNA (in green on figure 4) [43]. According to the ESDSA model, the single-stranded tail of a recessed fragment invades a partially overlapping fragment and primes DNA synthesis. However, contrary to what was observed during classical DNA replication of double-stranded DNA, newly synthesized single stranded DNA tails are generated by displacement of the D-loop in a manner similar to that observed during transcription, when newly-synthesized RNA is progressively liberated from the transcription machinery during the progression of RNA polymerase on the DNA template. The strand extension can proceed to the end of the DNA template. The newly synthesized single-stranded DNA tails generated by this process (in green in figure 4) can anneal if they contain complementary sequences thereby facilitating the precise reconstruction of long double-stranded DNA intermediates. These long intermediates subsequently recombine to reform a circular chromosome. Many enzymes play a key role in the ESDSA pathway, DNA polymerases Pol I and Pol III are required for the massive DNA synthesis step [44] whereas RecA is required for homologous strand invasion to prime DNA synthesis and for the maturation of the linear intermediates into full-size circular chromosomes through a classical recombination process. It was shown that ESDSA plays a major role in *D. radiodurans* radioresistance. A massive DNA synthesis beginning 2 h after γ-irradiation at a dose of 2,500 Gy was observed in *P. abyssi* [45] suggesting that ESDSA may also take place in Archaea. However, the presence of an efficient ESDSA mechanism in Archaea has not yet been investigated.

4.2.4. Non-homologous end-joining (NHEJ). It is the major pathway of DNA double-strand break repair in eukaryotes and it has only recently been identified and characterized in bacteria [46-48]. It was proposed that NHEJ could also take place in *D. radiodurans* because of the presence of proteins consistent with this process [49,50]. In particular, PprA, a protein specific from *Deinococcacea* and highly induced by ionizing radiation or desiccation [51] has been shown to preferentially bind to double-stranded DNA carrying strand breaks, to inhibit degradation of DNA by exonuclease, and to stimulate the DNA end-joining reaction catalyzed by DNA ligases [50,52]. Moreover, cells devoid of PprA are highly radiosensitive [50,51]. However, evidence for an active NHEJ process involved in DNA double-strand break repair in *D. radiodurans* remains to be discovered.

4.3. Processing of DNA ends

In Bacteria and Archaea, the three main pathways involved in DNA double-strand break repair require the formation of single-stranded DNA tails. This processing can be achieved through the activities of multiple helicases and nucleases (for review, see [53]). The *E. coli* RecBCD complex and AddAB, its *B. subtilis* functional analog, with their helicase and ATP-dependent nuclease activities play the major role in resection of DNA ends and in the loading of the RecA recombinase on SSB covered single-stranded DNA tails (figure 5A). However, if the RecBCD pathway is inactivated, an alternate pathway, the RecF pathway, promotes recombinational DNA double-strand break repair in *E. coli* [54-56]. The *E. coli* RecF pathway depends on the RecQ helicase and the RecJ exonuclease to resect duplex DNA, whereas RecO and RerR act with RecF to load RecA onto single-strand DNA covered by SSB [57] (figure 5A).

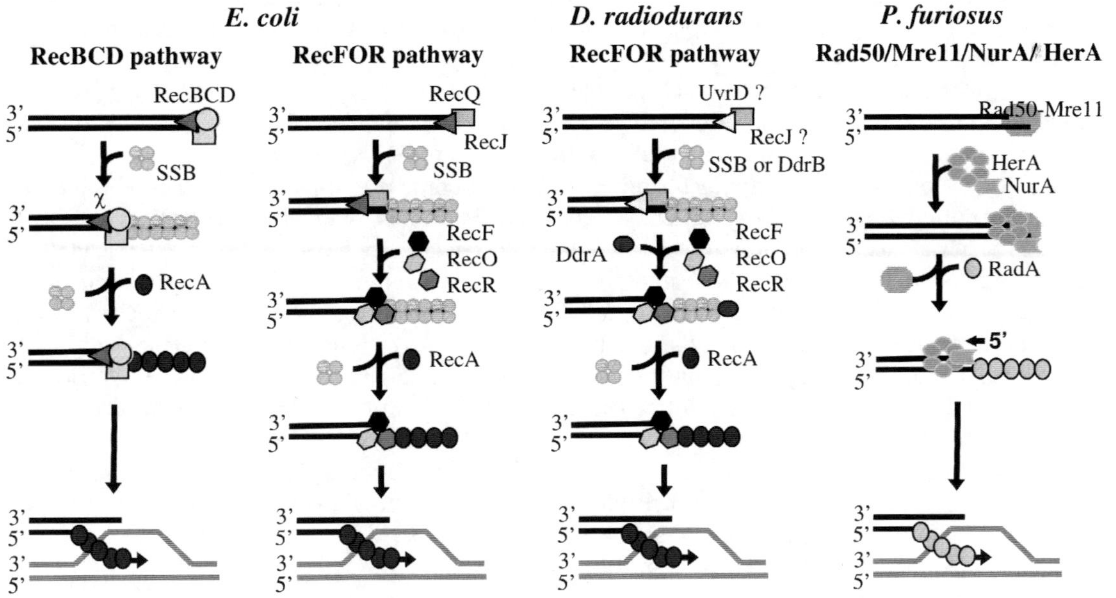

Figure 5. Main pathways involved in the processing of DNA ends.

D. radiodurans, like *D. geothermalis* and *D. deserti*, is naturally devoid of the RecB and RecC enzymes but encodes all the components of the alternate RecF pathway. We have recently shown that, in *D. radiodurans*, RecFOR proteins are essential for DNA double-strand break repair through ESDSA whereas RecJ protein is essential for cell viability [58]. In contrast, cells devoid of RecQ, the major helicase implicated in repair through the RecF pathway in *E. coli*, are resistant to γ-irradiation and display a wild type DNA double-strand break repair capacity. In contrast, *uvrD* mutants showed a markedly reduced radioresistance and a slow rate of fragment assembly suggesting that UvrD might be involved in the processing of double-stranded DNA ends (figure 5B).

Although Archaea are prokaryotes, most of the mechanisms and associated proteins involved in DNA metabolism (replication of DNA, transcription, translation, DNA repair and recombination pathways) are close to those found in eukaryotic cells. Eukaryotic cells use two main pathways to repair DNA double-strand breaks: HR and NHEJ depending on the phase of the cell cycle and the nature of the DNA ends. A key step where pathway choice is exerted is in the "licensing" of DNA ends to produce recombinogenic single-stranded tails. The conserved MRX/MRN complex, consisting of proteins Rad50, Mre11 and Xrs2/Nbs1 (in yeast /human) appears to be a major regulator of DNA end processing (for review, see [59]). Archaea encode proteins homologous to Rad50 and Mre11, but not to Xrs2/Nbs1. In hyperthermophilic archaea, Rad50 and Mre11 genes form a well conserved operon with two other genes encoding a nuclease (NurA) and a helicase (HerA/MlaA) [60-62]. *In vitro*, NurA and HerA proteins from *Pyrococcus furiosus* form a complex exhibiting helicase/nuclease activities stimulated by the presence of Rad50/Mre11 [63]. The Rad50/Mre11 complex generates short 3' ends before recruitment of the NurA/HerA complex. This allows degradation of the 5' end of DNA and generates a 3' protruding end which may be invaded by RadA (figure 5C). *In vivo* studies conducted in *Sulfolobus acidocaldarius* revealed that when cells were exposed to γ-rays at a dose of 1000 Gy, Mre11 was recruited on DNA [64]. HerA protein was also detected on the chromosome and formed a complex with Rad50 and Mre11. In contrast, NurA was not detected on the chromosome after ionizing irradiation and was not found associated within the Rad50/Mre11/HerA complex. Complementary approaches are required to elucidate *in vivo* the role of NurA in thermophilic Archaea DNA repair mechanism. The role of Rad50 and Mre11 was also analyzed *in vivo* in a polyploid archaeon *Haloferax volcanii*, which does not encode NurA and HerA proteins [65]. Unlike the yeast *Saccharomyces cerevisiae*, deletion of the *rad50* and *mre11* genes leads to a higher resistance of this archaeon to DNA damaging agents. In *H. volcanii*, homologous recombination is the primary pathway of DNA double strand break repair in cells devoid of Rad50 and Mre11 proteins, but not in the wild type where, as in mammalian cells, DNA double-strand break repair is mainly performed by NHEJ. However, the RadA protein remains essential for cell survival after DNA-damaging treatments [65]. RadA-dependent HR would probably be required late in the restoration of an intact circular chromosome, when most of the fragments are already reassembled through an NHEJ process. Further studies are required to identify the complete set of archaeal proteins involved in the NHEJ pathway.

As in bacteria, accessory proteins are required in eukaryotic cells to favor HR. Rad52 and Rad55/57 were shown to promote the loading of the Rad51 recombinase on DNA single-stranded tails covered by the SSB-like complex RPA in irradiated cells. Rad54 is a motor protein that translocate along double-stranded DNA, promotes chromatin remodeling and strongly stimulates the Rad51 DNA strand exchange activity. Together with *rad51* and *rad52*, *rad54* is one of the three most ionizing radiation sensitive single mutant in yeast *S. cerevisiae* [66]. In Archaea, proteins found by sequence homology with RadA may act as accessory proteins to promote HR. They include RadA2, RadB, RadC, KaiC/aRadC and Sms proteins (reviewed in [67]). The deletion of RadB in *H. volcanii* leads to a decrease of cell growth, a high UV sensitivity and recombination defects. *In vitro* studies of RadB from *P. furiosus* showed that the protein is able to interact with DNA, but does not catalyze the exchange of DNA strands required for homologous recombination. Other mediators may be involved in homologous recombination in Archaea, such as the homolog of the eukaryotic Rad54 protein. In *Sulfolobus solfataricus*, SsoRad54 is able to interact *in vitro* with DNA double strands, and was proposed to be a translocase. In addition, SsoRad54 also interacts with RadA protein to stimulate DNA strand exchange [68].

5. Nucleoid structure

Bacterial chromosomes form in association with proteins, condensed structures called nucleoids. The nucleoids of members of the radioresistant genera *Deinococcus* and *Rubrobacter* revealed a high degree of genome condensation and remain unaltered after high-dose γ-irradiation [69, 70]. Restricted diffusion of the DNA ends would be a reasonable explanation for efficient repair of the numerous DNA double-strand breaks generated by exposure to high doses of ionizing radiation. Moreover, it was proposed that the 4 to 10 copies of the *D. radiodurans* genome were prealigned [71], facilitating the search of homology to favor efficient error-free DNA double-strand break repair (HR and ESDSA). However, prealignment of the copies of homologous chromosomes has not been demonstrated.

The very abundant histone-like protein HU from *E. coli* is associated with the bacterial nucleoid and shown to be involved in cell survival after γ-irradiation [72]. In *D. radiodurans*, HU protein is essential for cell viability and, when expressed from a thermosensitive plasmid, its progressive cellular depletion at the non-permissive temperature generates decondensation and fractionation of the nucleoid preceding cell lysis, suggesting that the HU protein plays a major architectural role in the structure of the *D. radiodurans* nucleoid [73].

Archaea belonging to the *Euryarchaeota* lineage, such as *Pyrococcus* and *Thermococcus* species, as well as Methanogenic archaea, encode histone-like proteins displaying structural homology with those found in eukaryotes (reviewed in [74]). The archaeal nucleosome is not an octamer as in eukaryotes. In solution, archaeal histone-like proteins form dimers that can subsequently become tetramers to bind DNA. Several other small proteins have been characterized in some *Euryarchaea* species as the Methanogenic archaea that may possibly be involved in chromosome compaction [75, 76].

The chromatin structure of some species from the *Crenarchaeota* lineage, such as *Sulfolobus,* differs from those observed in *Pyrococcus* and *Thermococcus* archaea since to date, none of them have been found to encode histones. These organisms use at least two other DNA-binding proteins called AlbA and Sul7d to achieve DNA compaction (reviewed in [77]).

In eukaryotic cells, DNA repair requires local changes in chromatin structure to facilitate the accessibility of the lesions to the DNA repair machinery. Histone acetylases and histone desacetylases were shown to play a major role in the regulation of transcription by modulation of the compaction of the chromatin [78]. As in eukaryotes, there are several lines of evidence that Archaea could adjust chromatin compaction. In *Sulfolobus*, the DNA-binding affinity of AlbA is modulated by Pat and Sir2 enzymes through acetylation and deacetylation processes [79, 80]. Deacetylation of AlbA results in repression of transcription. The observed decrease of histones Htz1 and Htz2 concentration in *Thermococcus zilligii* when cells enters stationary phase cells may have a considerable effect on the overall structure of genomic DNA and consequently on gene expression [81]. Unfortunately, the data illustrating the global structure of archaeal genome during DNA repair remains scarce in the literature and the involvement of the nucleoid structure in archaeal radioresistance is still an open question.

6. Induced response to γ-irradiation

6.1. Transcriptome analysis

Analysis of the transcriptome of *D. radiodurans* revealed a novel group of genes that are up-regulated in response to either dessication or ionizing radiation [51]. Among the 72 genes up-regulated during the first hour after a sublethal dose of ionizing radiation, 33 are also highly expressed in *D. radiodurans* cultures recovering from desiccation. Only a limited number of well defined DNA repair genes including *recA* are found among these genes. The five genes whose expression is most highly induced in response to each stress include *ddrA*, *ddrB*, *ddrC*, *ddrD* and *pprA*. These genes are conserved in *D. geothermalis* and *D. deserti* but are not found outside of the *Deinococcus* lineage. Three of them were shown to play a major role in *D. radiodurans* radioresistance [51]. DdrA protects the single-stranded DNA ends against degradation [82] and is probably involved in protection of the recombinogenic DNA substrates [83] whereas DdrB binds to single-strand DNA [84] and PprA protects DNA double-stranded DNA ends against degradation and stimulates the activity of DNA ligases [50] suggesting a role of PprA in a putative NHEJ pathway. It is also interesting to note that the 3 predicted superoxide dismutases and the 3 predicted catalases which can protect biomolecules from ROS mediated damage are induced after exposure to ionizing radiation [51].

A transcriptional study was undertaken with *P. furiosus* exponentially growing cells irradiated with γ-Rays at a dose of 2500 Gy [33]. Under these conditions, 321 ORFs showed changes in gene expression in the time needed to repair all cell damage. In *P. furiosus*, the response to γ-irradiation affects various metabolic pathways as well as transcription, translation, various membrane transporters and also a large number of genes of unknown function. Interestingly, most of the systems involved in oxygen detoxification and redox homeostasis appeared to be highly and constitutively expressed. Moreover, the up-regulation of a gene encoding a ferritin/Dps-like protein in *P. furiosus* could trap free iron for limiting the production of hydroxyl radicals by Fenton chemistry [33]. The *radA*, *radB*, *dp1* and *dp2* genes encoding the two subunits of DNA polymerase D were shown to be slightly induced under these conditions but most genes known to be involved in DNA repair or in DNA replication do not show significant variation in gene expression. Unlike *P. furiosus*, RadA protein seems present at a constant level in *P. abyssi* after irradiation at 2,500 Gy [45]. One can assume that, due to the deleterious effects of temperature on DNA, repair genes are constitutively expressed in hyperthermophilic organisms.

Transcriptomic coupled with proteomic analysis was also performed in the halophilic archaeon *Halobacterium* sp. NRC-1 [85]. This analysis was conducted during 240 minutes post-irradiation incubation in cells exposed to 2500 Gy γ-irradiation. Genes encoding proteins involved in DNA repair (RadA recombinase, Hjr resolvase, UvrD helicase) or nucleotide biosynthesis (CTP synthase, cytidylate kinase) were found among the up-regulated genes. An inverse relationship was observed between mRNA changes of some genes involved in DNA replication and cell division suggesting that a pause occurred to ensure completion of DNA repair prior cell division. A significant increase of superoxide dismutase (SOD_2) and thioredoxin ($trxA_2$) levels was also observed after γ- irradiation to probably scavenge free radicals. Furthermore, eight dehydrogenases are down-regulated [85] probably to minimize ROS production by subsequent auto-oxidation reactions [26].

Two independent radioresistant derivatives of *Halobacterium* NRC-1, LH5 and LH7a, isolated by step doses of electron-beam radiation and selection of survivors [86], displayed a LD_{50} greater than 11,000 Gy. To understand how these mutants have acquired the ability to better cope with radiation, a transcriptomic analysis was performed to compare whole expression profile of the two mutants versus wild-type strain under standard growth conditions, [86]. The LH5 mutant showed 119 deregulated genes, 47 up-regulated and 72 down-regulated, compared to the wild strain. The genes encoding the RadA recombinase, the RPA protein complex (*rfa3*, *rfa8* and *ral* genes) and the superoxide dismutase enzymes (*sod1* and *sod2*) were shown to be induced, confirming their important role in the cellular DNA repair and detoxification. The LH7a mutant showed 47 deregulated genes including 11 that were up-regulated and 36 that were down-regulated. The comparison between LH5 and LH7a expression patterns showed that, only the 3genes encoding the trimeric RPA protein were induced in both independent mutants suggesting that the complex is critical for the radioresistance.

6.2. Regulatory network

In *E. coli*, DNA damaging treatments result in the induction of SOS functions, a variety of cellular processes mediated by the RecA-dependent cleavage of LexA, the repressor of SOS genes. *D. radiodurans* possesses two LexA homologs that undergo, as in *E. coli*, RecA-dependent cleavage after DNA damage. However, no regulon under the control of LexA1 or LexA2 proteins has been identified

to date in *D. radiodurans*. In particular, *recA* induction following γ-irradiation is not controlled by LexA1 or LexA2 but depends on a *Deinococcus* specific regulatory protein IrrE, also called PprI, that was shown to be a positive effector that enhances the expression of some DNA repair genes following exposure to radiation [50, 87, 88]. Recent crystallographic data combined with a site-directed mutagenesis study on the *D. deserti* IrrE protein, suggest that IrrE possesses a putative proteolytic activity essential for radiotolerance [89]. The presence of a common radiation/desiccation response motif (RDRM) found in the upstream region of *recA*, *pprA* and other highly radiation-induced genes in *D. radiodurans* and of their homologs in *D. geothermalis* and *D. deserti* suggests that a transcription regulator would target this common RDRM sequence [90, 91]. Direct binding of IrrE to the RDRM sequence has not, to our knowledge, been demonstrated. DdrO, another protein specific to *Deinococcaceae*, has been proposed to be this global regulator [91]. However, nothing is known about the precise role of IrrE and DdrO in the response to γ-irradiation or desiccation.
In Archaea, the transcription of several genes encoding general transcription factors, as well as regulators, was found to be modulated during cell recovery [33, 85]. These activation and repression are responsible of other gene regulation and coordinate the network of repair pathways to face cell injuries. Unfortunately, most of their targets and their hierarchical expression during repair have yet to be elucidated, thus limiting our understanding of the variety of archaeal responses to stress.

7. Conclusion

Different strategies were independently selected during evolution in Bacteria and Archaea to help cell to recover from radiation-induced damage. It is now generally accepted that the Archaea share many similarities in their information-processing pathways with Eukarya and could be used as simpler models of the much more complex eukaryal ones. Comparison of radioresistant bacteria and archaea has revealed several common strategies such as (i) limitation of ROS production or scavenging of free radicals (ii) efficient DNA double-strand break repair. However, other strategies are only found in some species (i) several radioresistant bacteria exhibit a low iron concentration to prevent production of $OH^{\cdot}$ but not *E. coli* radioresistant mutants (ii) the protection of *D. radiodurans* and *Halobacterium* proteins against oxidation plays a critical role for their ability to repair cell injuries (iii) several radioresistant bacteria and archaea, but not all, encode pigments (iv) many genes induced during post irradiation cell recovery, and involved in radioresistance, are specific for a particular phylum (v) *D. radiodurans* combines DNA double-strand break repair with massive DNA synthesis, also observed in *P. abyssi*, suggesting that ESDSA may also take place in Archaea. However, the presence of an efficient ESDSA mechanism in Archaea has not yet been investigated (vi) a condensed nucleoid structure could favor error-free genome reassembly by limiting the diffusion of DNA fragments. However, the proteins involved in the organization of the nucleoid are not the same in radioresistant bacteria and archaea. Radioresistance is the result of complex cell networks and further work is required to understand the combination of pathways that contributes to radioresistance at the molecular level.

Acknowledgments

F Confalonieri and S Sommer are supported by the Centre National de la Recherche Scientifique, the University Paris-Sud, and S. Sommer is supported by the Commissariat à l'Energie Atomique (CEA LRC42V), Electricité de France and the Agence Nationale de la Recherche (ANR-07-BLAN-0106, ANR-08-BLAN-0336-02). We thank all members of the groups "Radioresistance et Mutagenèse chez les *Deinococcacea*" and "Laboratoire de Génomique des Archaea" for their contribution, V. Favaudon and the Institut Curie for irradiation facilities and M. DuBow for brushing up the English.

References

[1] Rainey F A, Ray K, Ferreira M, Gatz B Z, Nobre M F, Bagaley D, Rash B A, Park M J, Earl A M, Shank N C, Small A M, Henk M C, Battista J R, Kampfer P and da Costa M S 2005 *Appl. Environ. Microbiol.* **71** 5225.

[2] Anderson A W, Nordon H C, Cain R F, Parrish G and Duggan D 1956 *Food Technol.* **10** 575.
[3] Battista J R and Rainey F A 2001 *Family I. Deinococcaceae* (New-York: Springer-Verlag)
[4] Callegan R P, Nobre M F, McTernan P M, Battista J R, Navarro-Gonzalez R, McKay C P, da Costa M S and Rainey F A 2008 *Int. J. Syst. Evol Microbiol.* **58** 1252.
[5] Cox M M and Battista J R 2005 *Nat. Rev. Microbiol.* **3** 882.
[6] Jolivet E, L'Haridon S, Corre E, Forterre P and Prieur D 2003 *Int. J. Syst. Evol. Microbiol.* **53** 847.
[7] Tapias A, Leplat C and Confalonieri F 2009 *Extremophiles* **13** 333.
[8] Shahmohammadi H R, Asgarani E, Terato H, Saito T, Ohyama Y, Gekko K, Yamamoto O and Ide H 1998 *J. Radiat. Res. (Tokyo)* **39** 251.
[9] Kottemann M, Kish A, Iloanusi C, Bjork S and DiRuggiero J 2005 *Extremophiles* **9** 219
[10] Harris D R, Pollock S V, Wood E A, Goiffon R J, Klingele A J, Cabot E L, Schackwitz W, Martin J, Eggington J, Durfee T J, Middle C M, Norton J E, Popelars M C, Li H, Klugman S A, Hamilton L L, Bane L B, Pennacchio L A, Albert T J, Perna N T, Cox M M and Battista J R 2009 *J. Bacteriol.* **191** 5240.
[11] Halliwell B and Gutteridge J M 1999 *Free radicals in biology and medecine* (Oxford: Oxford University Press)
[12] Gerard E, Jolivet E, Prieur D and Forterre P 2001 *Mol. Genet. Genomics* **266** 72.
[13] Blattner F R, Plunkett G, 3rd, Bloch C A, Perna N T, Burland V, Riley M, Collado-Vides J, Glasner J D, Rode C K, Mayhew G F, Gregor J, Davis N W, Kirkpatrick H A, Goeden M A, Rose D J, Mau B and Shao Y 1997 *Science* **277** 1453.
[14] Ng W V, Kennedy S P, Mahairas G G, Berquist B, Pan M, Shukla H D, Lasky S R, Baliga N S, Thorsson V, Sbrogna J, Swartzell S, Weir D, Hall J, Dahl T A, Welti R, Goo Y A, Leithauser B, Keller K, Cruz R, Danson M J, Hough D W, Maddocks D G, Jablonski P E, Krebs M P, Angevine C M, Dale H, Isenbarger T A, Peck R F, Pohlschroder M, Spudich J L, Jung K W, Alam M, Freitas T, Hou S, Daniels C J, Dennis P P, Omer A D, Ebhardt H, Lowe T M, Liang P, Riley M, Hood L and DasSarma S 2000 *Proc. Natl. Acad. Sci. U S A* **97** 12176.
[15] Robb F T, Maeder D L, Brown J R, DiRuggiero J, Stump M D, Yeh R K, Weiss R B and Dunn D M 2001 *Methods Enzymol.* **330** 134.
[16] White O, Eisen J A, Heidelberg J F, Hickey E K, Peterson J D, Dodson R J, Haft D H, Gwinn M L, Nelson W C, Richardson D L, Moffat K S, Qin H, Jiang L, Pamphile W, Crosby M, Shen M, Vamathevan J J, Lam P, McDonald L, Utterback T, Zalewski C, Makarova K S, Aravind L, Daly M J, Minton K W, Fleischmann R D, Ketchum K A, Nelson K E, Salzberg S, Smith H O, Venter J C and Fraser C M 1999 *Science* **286** 1571.
[17] Zivanovic Y, Armengaud J, Lagorce A, Leplat C, Guerin P, Dutertre M, Anthouard V, Forterre P, Wincker P and Confalonieri F 2009 *Genome Biol.* **10** R70.
[18] del Rio L A, Corpas F J, Sandalio L M, Palma J M, Gomez M and Barroso J B 2002 *J. Exp. Bot.* **53** 1255
[19] Omelchenko M V, Wolf Y I, Gaidamakova E K, Matrosova V Y, Vasilenko A, Zhai M, Daly M J, Koonin E V and Makarova K S 2005 *BMC Evol. Biol.* **5** 57.
[20] Brioukhanov A L, Netrusov A I and Eggen R I 2006 *Microbiology* **152** 1671.
[21] Grunden A M, Jenney F E, Jr., Ma K, Ji M, Weinberg M V and Adams M W 2005 *Appl. Environ. Microbiol.* **71** 1522.
[22] Jenney F E, Jr., Verhagen M F, Cui X and Adams M W 1999 *Science* **286** 306.
[23] Weinberg M V, Jenney F E, Jr., Cui X and Adams M W 2004 *J. Bacteriol.* **186** 7888.
[24] Molina-Heredia F P, Houee-Levin C, Berthomieu C, Touati D, Tremey E, Favaudon V, Adam V and Niviere V 2006 *Proc. Natl. Acad. Sci. U S A* **103** 14750
[25] Adam V, Royant A, Niviere V, Molina-Heredia F P and Bourgeois D 2004 *Structure* **12** 1729
[26] Imlay J A 2003 *Annu. Rev. Microbiol.* **57** 395.
[27] Archibald F S and Fridovich I 1982 *Arch. Biochem. Biophys.* **214** 452.

[28] Daly M J, Gaidamakova E K, Matrosova V Y, Vasilenko A, Zhai M, Leapman R D, Lai B, Ravel B, Li S M, Kemner K M and Fredrickson J K 2007 *PLoS Biol.* **5** e92.
[29] Daly M J, Gaidamakova E K, Matrosova V Y, Vasilenko A, Zhai M, Venkateswaran A, Hess M, Omelchenko M V, Kostandarithes H M, Makarova K S, Wackett L P, Fredrickson J K and Ghosal D 2004 *Science* **306** 1025.
[30] Seib K L, Tseng H J, McEwan A G, Apicella M A and Jennings M P 2004 *J. Infect. Dis.* **190** 136.
[31] Fredrickson J K, Li S M, Gaidamakova E K, Matrosova V Y, Zhai M, Sulloway H M, Scholten J C, Brown M G, Balkwill D L and Daly M J 2008 *Isme J* **2** 393.
[32] Kish A, Kirkali G, Robinson C, Rosenblatt R, Jaruga P, Dizdaroglu M and DiRuggiero J 2009 *Environ. Microbiol.* **11** 1066.
[33] Williams E, Lowe T M, Savas J and DiRuggiero J 2007 *Extremophiles* **11** 19.
[34] Krisko A and Radman M 2010 *Proc. Natl. Acad. Sci. U S A* **107** 14373.
[35] Krisko A, Smole Z, Debret G, Nikolic N and Radman M 2010 *J. Mol. Biol.* **402** 775.
[36] Bentchikou E, Servant P, Coste G and Sommer S 2007 *J. Bacteriol.* **189** 4784
[37] Hefferin M L and Tomkinson A E 2005 *DNA Repair (Amst)* **4** 639.
[38] Blasius M, Hubscher U and Sommer S 2008 *Crit. Rev. Biochem. Mol. Biol.* **43** 221.
[39] Wyman C and Kanaar R 2006 *Annu. Rev. Genet.* **40** 363.
[40] Paques F and Haber J E 1999 *Microbiol Mol Biol Rev* **63** 349.
[41] Seitz E M, Brockman J P, Sandler S J, Clark A J and Kowalczykowski S C 1998 *Genes Dev.* **12** 1248.
[42] Daly M J and Minton K W 1996 *J. Bacteriol.* **178** 4461.
[43] Zahradka K, Slade D, Bailone A, Sommer S, Averbeck D, Petranovic M, Lindner A B and Radman M 2006 *Nature* **443** 569.
[44] Slade D, Lindner A B, Paul G and Radman M 2009 *Cell* **136** 1044.
[45] Jolivet E, Matsunaga F, Ishino Y, Forterre P, Prieur D and Myllykallio H 2003 *J. Bacteriol.* **185** 3958.
[46] Chayot R, Montagne B, Mazel D and Ricchetti M *Proc. Natl. Acad. Sci. U S A* **107** 2141.
[47] Shuman S and Glickman M S 2007 *Nat Rev Microbiol* **5** 852.
[48] Weller G R, Kysela B, Roy R, Tonkin L M, Scanlan E, Della M, Devine S K, Day J P, Wilkinson A, d'Adda di Fagagna F, Devine K M, Bowater R P, Jeggo P A, Jackson S P and Doherty A J 2002 *Science* **297** 1686.
[49] Lecointe F, Shevelev I V, Bailone A, Sommer S and Hubscher U 2004 *Mol.. Microbiol.* **53** 1721.
[50] Narumi I, Satoh K, Cui S, Funayama T, Kitayama S and Watanabe H 2004 *Mol. Microbiol.* **54** 278.
[51] Tanaka M, Earl A M, Howell H A, Park M J, Eisen J A, Peterson S N and Battista J R 2004 *Genetics* **168** 21.
[52] Murakami M, Narumi I, Satoh K, Furukawa A and Hayata I 2006 *Biochim. Biophys. Acta* **1764** 20.
[53] Yeeles J T and Dillingham M S *DNA Repair (Amst)* **9** 276.
[54] Horii Z and Clark A J 1973 *J. Mol. Biol.* **80** 327.
[55] Mahdi A A and Lloyd R G 1989 *Mol. Gen. Genet.* **216** 503.
[56] Sakai A and Cox M M 2009 *J. Biol. Chem.* **284** 3264.
[57] Handa N, Morimatsu K, Lovett S T and Kowalczykowski S C 2009 *Genes Dev* **23** 1234.
[58] Bentchikou E, Servant P, Coste G and Sommer S 2010 *PLoS Genet* **6** e1000774.
[59] Mimitou E P and Symington L S 2009 *DNA Repair (Amst)* **8** 983.
[60] Constantinesco F, Forterre P and Elie C 2002 *EMBO Rep.* **3** 537.
[61] Constantinesco F, Forterre P, Koonin E V, Aravind L and Elie C 2004 *Nucleic Acids Res.* **32** 1439.

[62] Manzan A, Pfeiffer G, Hefferin M L, Lang C E, Carney J P and Hopfner K P 2004 *EMBO Rep.* **5** 54.
[63] Hopkins B B and Paull T T 2008 *Cell* **135** 250.
[64] Quaiser A, Constantinesco F, White M, Forterre P and Elie C 2008 *BMC Mol Biol* **9** in press.
[65] Delmas S, Shunburne L, Ngo H P and Allers T 2009 *PLoS Genet.* **5** e1000552.
[66] Mazin A V, Mazina O M, Bugreev D V and Rossi M J *DNA Repair (Amst)* **9** 286.
[67] Haldenby S, White M F and Allers T 2009 *Biochem. Soc. Trans.* **37** 102.
[68] Haseltine C A and Kowalczykowski S C 2009 *Nucleic Acids Res.* **37** 2757.
[69] Levin-Zaidman S, Englander J, Shimoni E, Sharma A K, Minton K W and Minsky A 2003 *Science* **299** 254.
[70] Zimmerman J M and Battista J R 2005 *BMC Microbiol.* **5** 17.
[71] Minton K W 1994 *Mol. Microbiol.* **13** 9.
[72] Boubrik F and Rouviere-Yaniv J 1995 *Proc. Natl. Acad. Sci. U S A* **92** 3958.
[73] Nguyen H H, de la Tour C B, Toueille M, Vannier F, Sommer S and Servant P 2009 *Mol. Microbiol.* **73** 240.
[74] Sandman K and Reeve J N 2006 *Curr Opin Microbiol* **9** 520.
[75] Pavlov N A, Cherny D I, Nazimov I V, Slesarev A I and Subramaniam V 2002 *Nucleic Acids Res* **30** 685.
[76] Teyssier C, Toulme F, Touzel J P, Gervais A, Maurizot J C and Culard F 1996 *Biochemistry* **35** 7954.
[77] White M F and Bell S D 2002 *Trends Genet.* **18** 621.
[78] Hager G L, McNally J G and Misteli T 2009 *Mol. Cell* **35** 741.
[79] Bell S D, Botting C H, Wardleworth B N, Jackson S P and White M F 2002 *Science* **296** 148.
[80] Marsh V L, Peak-Chew S Y and Bell S D 2005 *J. Biol. Chem.* **280** 21122.
[81] Dinger M E, Baillie G J and Musgrave D R 2000 *Mol. Microbiol.* **36** 876.
[82] Harris D R, Tanaka M, Saveliev S V, Jolivet E, Earl A M, Cox M M and Battista J R 2004 *PLoS Biol.* **2** e304.
[83] Jolivet E, Lecointe F, Coste G, Satoh K, Narumi I, Bailone A and Sommer S 2006 *Mol. Microbiol..* **59** 338.
[84] Norais C A, Chitteni-Pattu S, Wood E A, Inman R B and Cox M M 2009 *J. Biol. Chem.* **284** 21402.
[85] Whitehead K, Kish A, Pan M, Kaur A, Reiss D J, King N, Hohmann L, DiRuggiero J and Baliga N S 2006 *Mol Syst Biol* **2** 47.
[86] DeVeaux L C, Muller J A, Smith J, Petrisko J, Wells D P and DasSarma S 2007 *Radiat. Res.* **168** 507.
[87] Earl A M, Mohundro M M, Mian I S and Battista J R 2002 *J. Bacteriol.* **184** 6216.
[88] Hua Y, Narumi I, Gao G, Tian B, Satoh K, Kitayama S and Shen B 2003 *Biochem. Biophys. Res. Commun.* **306** 354.
[89] Vujicic-Zagar A, Dulermo R, Le Gorrec M, Vannier F, Servant P, Sommer S, de Groot A and Serre L 2009 *J. Mol. Bio.l* **386** 704.
[90] de Groot A, Dulermo R, Ortet P, Blanchard L, Guerin P, Fernandez B, Vacherie B, Dossat C, Jolivet E, Siguier P, Chandler M, Barakat M, Dedieu A, Barbe V, Heulin T, Sommer S, Achouak W and Armengaud J 2009 *PLoS Genet.* **5** e1000434.
[91] Makarova K S, Omelchenko M V, Gaidamakova E K, Matrosova V Y, Vasilenko A, Zhai M, Lapidus A, Copeland A, Kim E, Land M, Mavrommatis K, Pitluck S, Richardson P M, Detter C, Brettin T, Saunders E, Lai B, Ravel B, Kemner K M, Wolf Y I, Sorokin A, Gerasimova A V, Gelfand M S, Fredrickson J K, Koonin E V and Daly M J 2007 *PLoS ONE* **2** e955.

Journal of Physics: Conference Series **261** (2011) 012006 doi:10.1088/1742-6596/261/1/012006

Synergy between low and high energy radical femtochemistry

YA Gauduel
Laboratoire d'Optique Appliquée, CNRS UMR7639, Ecole Polytechnique ParisTech - ENS Techniques Avancées ParisTech, 91761 Palaiseau Cedex, France and *Radiation Biology* MELUSYN Network, Armir-Collège de Polytechnique, Paris, France.

E-mail: yann.gauduel@ensta-paristech.fr

Abstract. The deleterious effects of ionizing radiation on integrated biological targets being dependent on the spatio-temporal distribution of short-lived radical processes, a thorough knowledge of these early events requires a real-time probing in the range $10^{-15} - 10^{-10}$ s. This manuscript review is focused on the synergy that exists between low (1-10 eV) and high (MeV) energy radiation femtochemistry (LERF, HERF respectively). The synergy remains crucial for the investigation of primary radical processes that take place within the prethermal regime of low energy secondary electrons. The quantum character of very-short lived electron in a prehydrated configuration provides a unique sub-nanometric probe to spatially explore some early radiation-induced biomolecular damage. This approach would foreshadow the development of innovative applications for spatio-temporal radiation biology such as, [i)] a highly-selective pro-drug activation using well-defined quantum states of short-lived radicals, [ii)] the real-time nanodosimetry in biologically relevant environments, and [iii)] the ultrashort irradiation of living cells.

1. Introduction

It is commonly admitted that the spatial distribution of radicals can be decisive for the induction and amplification of early biomolecular damage induced by different types of radiations (X and UV photons, relativistic particles, accelerated ions) [1-6]. From a fundamental point of view, one challenge concerns the complete understanding of multiscale events triggered by an initial energy deposition inside confined clusters of ionization and evolving over several orders of magnitude, typically from femtosecond (1fs = 10^{-15}s) and sub-micrometric scales. In condensed phase, the investigation of elementary radical reactions can be carried out in strong synergy with the most recent progresses of ultra-short laser sources and low or high energy radiation femtochemistry [6-8].

One of the most fundamental aspects of biomolecular damage triggered by radiations concerns the dissociative electron attachment processes that take place in confined ionization spaces. Such processes involve a hierarchy of secondary electron populations for which the energy varies from the thermal value (kT ~ 0.025 eV) to the sub-excitation and relativistic levels i.e. few eV, MeV respectively (figure 1). In presence of water molecules, the solvent of life, the primary radical processes induced by an electron attachment can be assisted or impeded by ultrafast caging process such as solvation phenomena, solvent bridged pair formation. The microscopic understanding of primary radical processes that are triggered by ionizing radiation requires the real-time probing of multiple non-equilibrium states whose the lifetimes are typically in the sub-picosecond regime. In this way, prethermal processes and sub-structure of tracks can play a crucial role at early time.

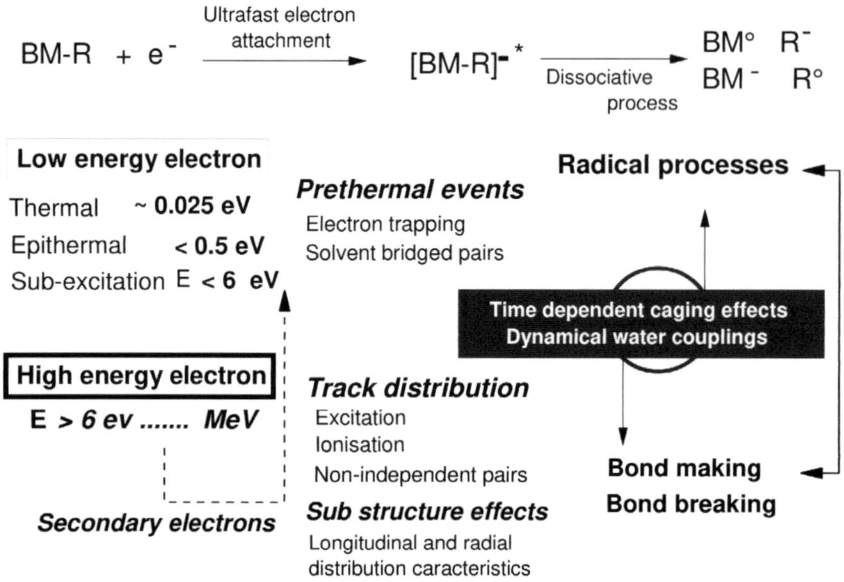

Figure 1. Relationship between low and high energy electrons involved in ultrafast electron attachment and dissociative process on biomolecules (BM). In aqueous environment, time dependent caging effects assist dissociative processes (bond making-bond breaking) and early radical events.

2. Low energy femtoradical chemistry in condensed phase

With the intensive development of ultra-short laser technologies and advanced high-time resolved spectroscopic methods, the course of short lived non-equilibrium radical trajectories are more and more observable on the molecular motion scale [7].

2.1 LERF of water molecules

In the framework of non-linear interactions of pure water molecules with ultra-short high energy UV pulses whose the peak power density is around 10^{10} W.cm^{-2}, a direct excitation can be triggered by a two-photon process. For wave plane propagation through an aqueous sample, this phenomenon is expressed by the equation 1:

$$\partial I/\partial x + \left(\frac{1}{v}\right)\partial I/\partial t = -BI^2 \qquad (1)$$

for which I is the radiation intensity, B the two photon absorption coefficient and v the light velocity in the medium.

Considering an excited $\tilde{A}$ state (1b1 $\rightarrow$ 3a1 for instance), a nonlinear two-photon UV excitation process (E_{Excit} = 2 x 4 eV) can be used to investigate the LERF signals dynamics of early water molecules radical processes. Femtosecond UV-IR absorption spectroscopic investigations in the energy range 3 - 1 eV allow to discriminate the sequential events of multiple radical channels (figure 2). The non-linear two-photon energy deposition of 8 eV in a water bulk triggers early water defects such as the hydronium ion and hydroxyl radical via an ultrafast positive hole $H_2O^{\bullet+}$ reaction, multiple

Journal of Physics: Conference Series **261** (2011) 012006 doi:10.1088/1742-6596/261/1/012006

non-equilibrium delocalized electron configurations (quasi-free delocalized electron $\{e_{-qf}\}$, p-like excited prehydrated electron $\{e^-_p\}$, electron-radical pairs and hydrated electron ground state$\{e^-_s\}$). All these transient radical pathways take place in less than 5×10^{-13} s.

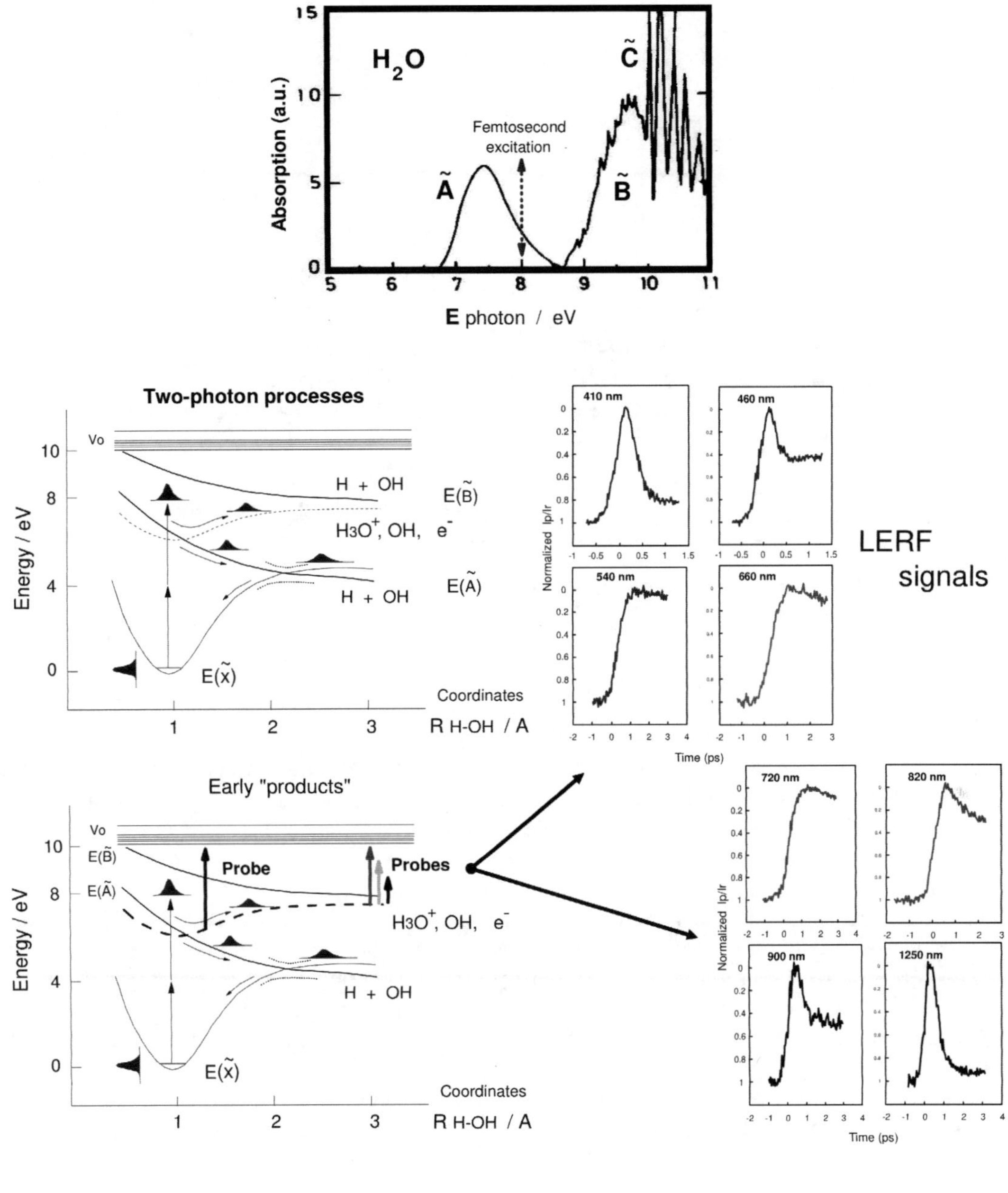

Figure 2. LERF of water molecules. Upper part: Electronic absorption spectrum of the first and second excited states of water molecules in the UV region. Middle and lower parts: Energy level diagrams of femtosecond two-photon excitation of water molecules leading to multiple radical pathways (early products). Some LERF signals probed at different wavelengths, in the UV-IR spectral range, are reported on the right part. Adapted from [7 and 10].

A direct electron photodetachment in the vicinity of water conduction band takes place in less than 250 fs. This primary phenomenon leads to an electron hydration via the relaxation of infrared 2p-like excited prehydrated electron: $2p(e^-_{prehyd}) \rightarrow 1s(e^-_{hyd})$ transition [12-13]. LERF experiments have also shown that an ultrafast pathway contributes to transient solvent bridged three-bodies complex $[OH^{\bullet}\ldots e^-\ldots H_3O^+]$ (figure 3).

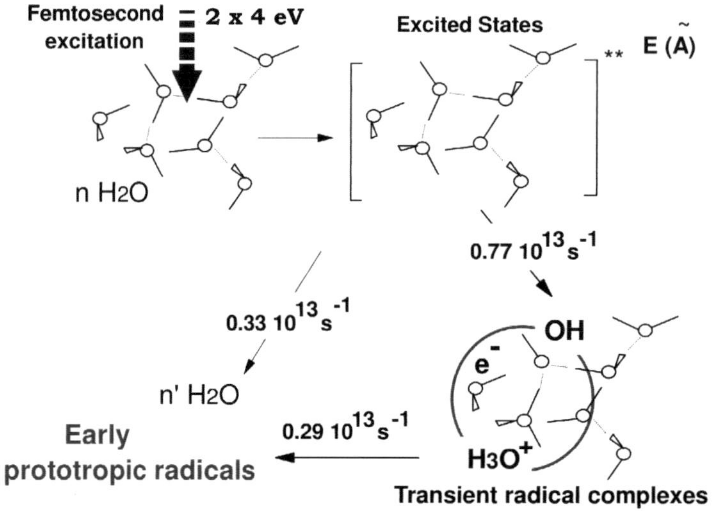

Figure 3. LERF of short-lived three-bodies complex $[OH^{\bullet}\ldots e^-\ldots H_3O^+]$ following a femtosecond two-photon UV excitation of water molecules in liquid phase. Adapted from [7,10].

The LERF of transient radical complexes supports the conclusion that $[H_3O^{+\cdots}e^-..OH]_{nH2O}$ deactivation ($0.29 \times 10^{13}\ s^{-1}$) is comparable to the estimate of a vibrationally excited water molecules relaxation ($\nu_{H2O}^* \sim 0.33 \times 10^{13}\ s^{-1}$) [14]. Indeed, LERF of neat liquid water contribute to deeply understand $^{i)}$ the nature of early branching between ultrafast radical pathways, $^{ii)}$ the contribution of ultrashort-lived solvent configurations to multiple potential energy surface crossing zones [15].

2.2 Ultrafast protic radical events in well-defined configurations

A fundamental question about early radical reactions in solution concerns the role of microscopic solvent dynamics during the reorganization of charge distribution. The understanding of solvent effects involved in the definition of reaction dynamic factors encompasses the synergy between high-time resolution of short-lived transition states and advances in computational solution chemistry [16-24]. Aqueous ionic solutions represent complicated many body systems and a paradigm for the investigation of elementary radical processes for which solvent caging effects can assist or impede reactive modes. An archetype concerns a dehydrogenation process for which a concerted electron-proton transfer can be considered as a transient state of the reaction pathway (equation 2).

$$\begin{pmatrix} RH \\ + \\ A^- \end{pmatrix}_S \xrightarrow{Excitation} \left| \begin{matrix} RH \\ \updownarrow \\ A^- \end{matrix} \right|^*_S \xrightarrow{Ionization} \left\langle \begin{matrix} RH\ldots e^- \\ A.^{\cdot} \end{matrix} \right\rangle^*_S \Rightarrow \left\langle \begin{matrix} R^- \\ \vdots \\ H^+, e^- \\ \vdots \\ A \end{matrix} \right\rangle^*_S \xrightarrow{T.E.} \begin{pmatrix} R^{-\cdot} \\ \vdots \\ H \\ \updownarrow \\ A \end{pmatrix}_S \xrightarrow{Rel} \begin{Bmatrix} R^{\cdot -} \\ AH \end{Bmatrix}_S \qquad (2)$$

The protic nature of water molecules may influence the quantum aspects of unbound/bound electron states, mainly during short-time concerted electron-proton transfer [25-30]. Indeed, femtosecond IR spectroscopy of elementary electron transfers in acid solutions offers the opportunity to investigate different early partitions between reactive and nonreactive pathways (figure 4). In concentrated hydrochloric acid solution (molecular ratio $H_2O/HCl = 5$), water molecules are shared by chloride ions and hydronium ions; consequently the Grotthuss mechanism is totally abolished [31]. With such conditions an ultrafast 2p-state electron transfer reaction with an aqueous proton (hydronium ion) is not observed. However, when the delocalised electron is trapped ~ 1 eV below the p-like electron level, an efficient prehydration electron-proton reaction takes place via a short-lived three-body complex ($Y...e-...H^+$) with Y=Cl. Femtosecond investigations have established that a well defined hydrogen bond ring structure $\{Cl..e^-..H^+\}_{aq}$ connecting a halide ion with an hydrated proton favors an efficient electron-proton reaction with a frequency rate of 4×10^{12} s^{-1} (figure 4).

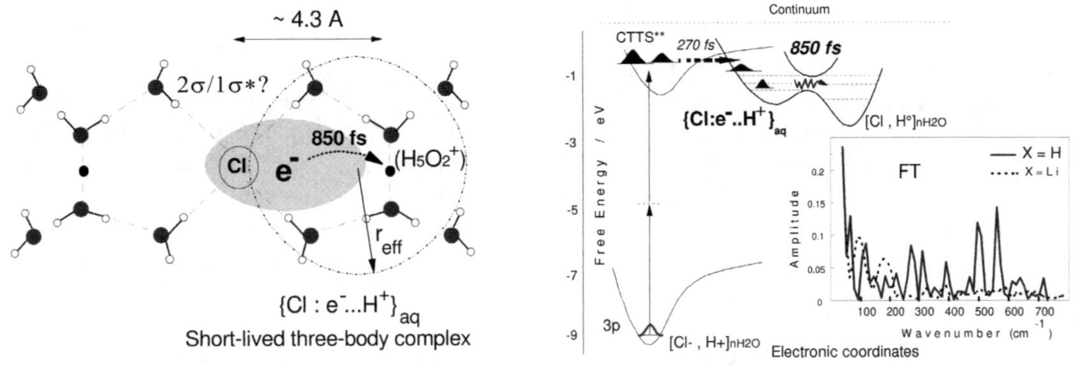

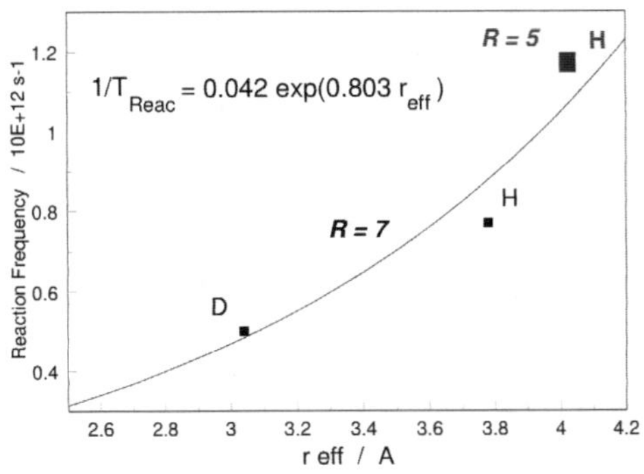

Figure 4. Ultrafast electron transfer trajectory inside the spherical volume of well-coordinated $[Cl^-...n\ H_2O...H^+]$ complex. Water caging effects characterized by specific vibrationnal modes assist the electron trajectory form excited CTTS states of chloride ion (excited Charge Transfer to Solvent State). The effective radius of this electron transfert is dependent on isotopic substitution and R value of the aqueous ionic solution [11].

This ultrafast electron-proton reaction is mainly governed by a transition state for which a delocalized electron oscillates between a polarizable Cl atom and the dihydronium ion, before an electron collapse with the hydrated proton. Considering semi-quantum MD simulations, the 4s character of a delocalised electron in the vicinity of Cl atom would be more favourable for an ultrafast electron transfer that the 2p-state excited electron [32]. The potential energy surface crossing zone would correspond to an "inverted Marcus region" [33]. The discrimination of vibrational modes during the formation of the pre-reactive three-body complex ($Y...e-...H^+$) argues for the contribution of selective solvent modes such as a vibration of torsional type or stretching mode of hydrogen bridge OH...O. The discrimination of high frequency modes around 65 cm^{-1} is assigned to the flexion of O-H...O within a hydrogen bond network. Consequently, a solvent caging effect involving vibronic couplings would assist an electronic migration on the potential energy surface of the hydrated proton.

The LERF investigations of early radical processes occurring in aqueous environments provide guidance for understanding the role of solvent-bridged electron-radical complexes in aqueous clusters of ionization triggered by high energy radiations.

3. High energy radiation femtochemistry of nascent tracks

The wide impact of ionizing radiation for biomedical applications (radiation biology and cancer radiotherapy) concerns the interactions of high-energy particles or photons (electron, proton, X and γ rays) with complex biomolecular architectures. Such interactions induce electronic or vibrational excitations, ionizations, ultrafast radical phenomena and primary molecular damage such as DNA strand breaks. These early radiation events take place in confined spaces, commonly called tracks or spurs. Numerous pulsed radiolysis experiments performed from microsecond to picosecond scales have greatly enhanced our understanding of water radiation chemistry and radical processes in tracks, considering nonhomogeneous and homogeneous regimes [4,5,34-41]. In nascent spurs, the escape probability of secondary low energy electrons would be influenced by ultra-fast recombination with two neoformed OH, H_3O^+ entities. These primary radical events would influence the yield of water decomposition products at early times. Up to now the indirect determination of the primary yield of fully relaxed hydrated electrons, a ubiquitous reducing radical stabilized by water molecules and whose the interaction potential with water bath is around -2.8 V [42-44], has been the subject of numerous investigations using linear accelerators. The quantum character of this molecular hydrogen precursor (1s-like ground state) [45-49] greatly determines the behavior of multiple radical and redox reactions in irradiated aqueous environments. Consequently, the early fate of hydrated electrons in nascent spurs is of particular relevance but has never been observed at early time after an energy deposition.

In this context, some recent innovative high-energy laser techniques open new opportunities for the development of femtosecond investigations of low-energy electron fate in radiation-induced aqueous nascent tracks. With the advent of powerful laser systems (TW laser), laser plasma interactions [8,9,50-54] can now provide high-energy, femtosecond electron bunches and allow the real-time probing of ultrafast radical events in nascent tracks (figure 5).

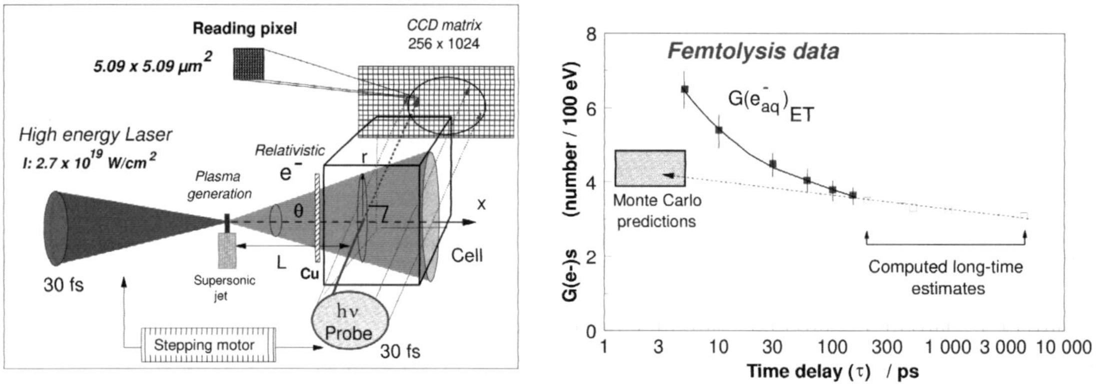

Figure 5. Left: Experimental set-up used for the development of HERF in aqueous liquid phase [9,55]. The energy of the femtosecond relativistic electron bunches is in the range 3-15 MeV. The Femtolysis detection uses a perpendicular pump (electron bunch) – probe (optical pulse) configuration. Right: Calculated estimate of the yield of hydrated electrons at early time G $(e^-_{aq})_{ET}$ from femtolysis water experiments performed in the temporal range 0-150 ps. The estimated $G(e^-_{aq})_{ET}$ value at $t_{Smax} \sim 5$ ps equals 6.5 ± 0.5 molecules / 100 eV.

FEMTOLYSIS (FEMTOsecond radioLYSIS) of pure liquid water has been performed by using an innovative laser produced high-energy, ultra-short electron bunches in the 2.5-15 MeV range and High Energy Radiation Femtochemistry (HERF) measurements [55]. The short-time monitoring of a primary reducing radical, hydrated electron e^-_{aq}, has been performed in confined ionization spaces (nascent spurs). The calculated yield of hydrated electrons at early time, $G(e^-_{aq})_{ET}$, is estimated to be 6.5 ± 0.5 (number/100 eV) at t ~ 5 ps after the ultrafast energy deposition [55]. This estimated value is high compare to [i)] the available data of previous works that used scavenging techniques, [ii)] the predictions of stochastic water radiolysis modelling for which the initial behaviour of hydrated electron is investigated in the framework of a classical diffusion regime of independent pairs. The HERF data give new insights into the early ubiquitous radical escape probability in nascent aqueous spurs and emphasize the importance of short-lived solvent bridged electron-radical complexes $[H_3O^{+\cdots}e_{aq}^-..OH]_{nH2O}$ *(non-independent pairs)*. A complete understanding of the $G(e^-_{aq})_{ET}$ value needs to account for quantum aspects of 1s-like trapped electron ground state and neoformed prototropic radicals that govern ultra-fast recombination processes via non-independent pair configurations. Indeed, the predicted estimate of $G(e^-_{aq})_{ET}$ value at $t \leq 5$ ps would be compared with the predictions of future quantum mechanical theories on ultrafast electron-radical pair dynamics in nascent spurs.

These FEMTOLYSIS data would request the semi-quantum modeling of early spatial secondary electron trajectories, considering the kinetic energy scattering and localization process. Further HERF researches will more deeply investigate several fundamental aspects such as the complex branching between ultra-fast excitation and ionization pathways in tracks - the relationship between sub-picosecond electron prehydration and ultrafast formation dynamics of H_2O^+, OH or H_3O^+ within non-independent radical pairs - the initial 3D spatio-temporal distribution of primary radical products inside heterogenous structures of native tracks. Regarding [i)] the recent developments of low energy bioradical femtochemistry of water molecules for which the "2p-like prehydrated electron" may exert a major role in biomolecular damage [57-60], [ii)] the advances in modules of the physical-biophysical Monte Carlo track structure codes [39] and [iii)] the quantum chemistry of radiations interactions with biomolecules [39], we should wonder whether ultrafast radiation damage occurring in the prethermal regime of secondary electrons represent a tenuous frontier between direct and indirect effects.

4. LERF and HERF of prethermal radical processes

Ultrafast radical events occurring in less than 1ps (10^{-12}s) after ionizing radiation deposition represent a specific domain for which the quantum character of short-lived events is preeminent. This is particularly important when biomolecular damage take place in the prethermal regime of secondary electrons (figure 1). Following a local energy loss around 20 eV deposited by a relativistic particle on a distance shorter than 100 Å, the secondary electrons scatter progressively their excess energy within a sub-micrometric environment. In the specific prethermal regime, the partially localized electron energy is still higher than the thermal energy kT i.e. E ~ 0.025 eV. As the primary biological target damages induced by ionizing radiations are highly dependent on the survival probability of low energy secondary electrons, a thorough knowledge of their short-range interactions with neighboring molecules requires the real-time probing of early events inside nanometric clusters of ionization.

In the framework of an energy/space/energy profile, the prethermal regime of localized electrons corresponds to the specific temporal window 0.1-10 ps. A major stride of biological interest concern the ultrafast radical processes mediated by p-like excited prehydrated electron. The LERF of biomimetic organized assemblies containing oxidized pyridine nucleotides (NAD^+, NMN^+) [61] or disulfide biomolecules [57] has demonstrated that very short-lived localized electrons contribute to ultrafast radical processes (figure 6).

During the prethermal regime, the activate barrier of a reduction process involves a crossing zone between a subpicosecond radiationless p $\rightarrow$ s transition and an efficient "p state" excited electron attachment on the biomolecule. It has been clearly established that a nondiffusive electron transfer involving an IR p-like electron is about *4.4 10^2 time faster* than a diffusion-controlled reaction with a hydrated electron ground state (1s state) [57,62,63]. Considering the similarity of pyridine nucleotides with DNA components, the pioneering LERF works on ultrafast pyridinyl radical processes induce an open question: Do prehydrated electron trigger direct DNA damage and mutagenic effect? Recent works performed under UV/ionizing radiation have established that femtosecond one-electron reductions of 5-halo-2'-deoxyuridines (potential radiosensitizing agents increasing the biological effects of radiations) and other nucleotides could be induced by prehydrated electron [58,63,64].

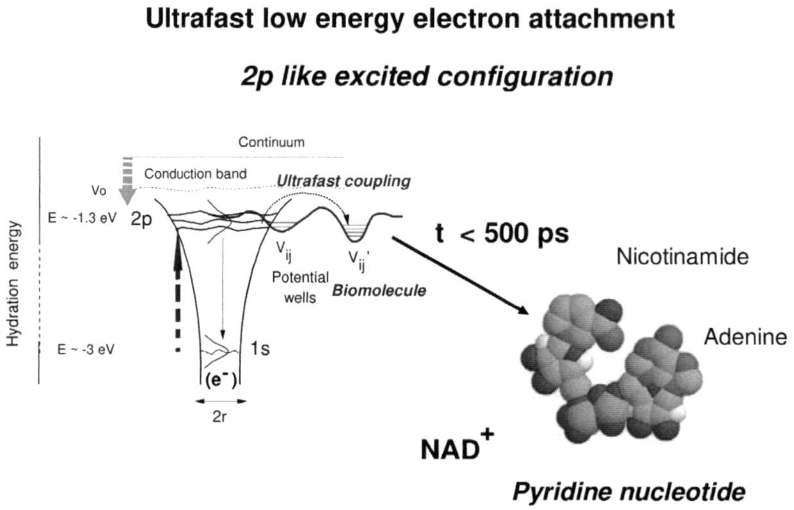

Figure 6. Energy level diagram of an ultrafast one-electron attachment on oxidized pyridine nucleotide (NAD^+) in aqueous environment. This process involves a 2p-like excited electron state and leads to the formation of a pyridinyl radical.

During the last decade, the LERF of prethermal radical processes has also been conducted with disulfide biomolecules whose cystamine, the reactive center of oxidized glutathione [57]. The interest of this biological disulfide is two-fold: the limited number of atoms for real-time investigations and clinical properties in living systems [66]. The direct attachment of a delocalized electron on a disulfide biomolecule has been observed in real time by femtosecond IR spectroscopy of homogeneous solutions or organized assemblies. The early partition between reactive ($\{e^-_{IR}\}_{Reac}$) and nonreactive ($\{e^-_{IR}\}_{p\text{->}s}$) electronic configurations could be determined from the following expression:

$$\frac{\alpha^{\omega T}_{\{e-IR\}\,\mathrm{Re}ac}}{\alpha^{\omega T}_{\{e-IR\}\,p->s}} = \frac{\sigma^{\omega T}_{\{e-IR\}\,\mathrm{Re}ac}\cdot P_{\mathrm{Re}ac}}{\sigma^{\omega T}_{\{e-IR\}\,p->}\cdot P_{p->s}} \tag{3}$$

In this expression, $P_{\mathrm{Re}ac}, P_{p->s}$ represent respectively the probability of a prehydration univalent reduction form a 2p-like excited electron and an electron salvation process via a 2p→1s transition. The probability of a prehydration univalent reduction $P_{\mathrm{Re}ac}$ is defined as follows:

$$\frac{P_{\mathrm{Re}ac}}{P_{p->s}} = \frac{\alpha^{\omega T}_{\{e-IR\}\,\mathrm{Re}ac}}{\alpha^{\omega T}_{\{e-IR\}\,p->s}} \tag{4}$$

$$P_{\mathrm{Re}ac} = 1 - P_{p->s} = 1 - \frac{\alpha^{\omega T}_{\{e-IR\}p->s}}{\left(\alpha^{\omega T}_{\{e-IR\}\mathrm{Re}ac} + \alpha^{\omega T}_{\{e-IR\}p->s}\right)} \quad (5)$$

In equations 3-5, the adjusted parameter $\alpha_i^{\omega T}$ represents the spectral contribution of an electronic state « i » at 0.99eV. The subpicosecond competition involving the ultrafast electron attachment on cystamine and the energy gap relaxation from the first excited state to the ground state of trapped electron is characterized by a high branching ratio ($P_{Reac}/ P_{p->s} \sim 0.91$). This ultrafast radical process would involve some distorsions of p-like orbitals of prehydrated electron in the vicinity of a sulfur-sulfur bond. The sub-picosecond lifetime of IR (e^-_p) represents a temporal limit for that an early SS bond energy rearrangement facilitates an electron tunnelling through a potential barrier separating the IR p-state electron of an unpaired electron. In this way, the calculated effective radius r_{eff} of a cystamine molecule involved in the non-equilibrium p-like excited electron attachment equals 9.8 Å. Regarding the native $RS{\therefore}SR^-$ radical anion, the antibonding electron contributes to the formation of a 2center-3 electron bond in less than 600 fs [57,67].

In synergy with theseLERF investigations, the early radiation-induced biomolecular target damage in nascent ionization tracks are been recently investigated by HERF, using ultra-short relativistic electron beams in the energy range 3-15 MeV and ultrafast spectroscopic methods. Following a dose of 15 Gy delivered in less than 500 fs, an ultrafast collapse takes place between low energy electron (precursors of fully hydrated electrons) and cystamine molecules localized in the chemical core i.e. at the interface between the physical core and penumbra zone. The early quantum yield of fully hydrated electrons in native tracks argues for a prethermal one-electron attachment on a disulfide biomolecule (cystamine) in less than 0.85×10^{-12} s. Consequently, the anionic radical formation of cystamine $(RS{\therefore}SR^-)_{aq}$ with $R = NH_2\ (CH_2)_2$ occurs faster than the radiationless $2p \rightarrow 1s$ transition of prehydrated electrons. In the future, predictive ionizing radiation effects on more complex biomolecular systems would be obtained in the framework of spatio-temporal correlations of energy deposition. In this way, disulfide molecules can be considered as a sensor of ultrafast oxidation-reduction phenomena involving sulfur-centered radical anion ($RS{\therefore}SR^-$). Time-resolved spectroscopic studies of a $2\sigma/1\sigma^*$ sulfur-sulfur bond dynamics from thiol monomers or disulfide biomolecule open a new era for sulfur biochemistry and provide guidance for further theoretical investigations of short-time solvent cage effect during a $2\sigma/1\sigma^*$ sulfur-sulfur bond breaking.

5. Future challenges and concluding remarks

The emerging HERF domain provides guidance for resolving some fundamental aspects of high energy radiation induced radical events during the nonhomogeneous regime of nascent tracks (figure 7). In this way, the synergy between LERF and HERF suggests that the quantum characters of trapped electron and ultrashort-lived water cation would greatly influence the dynamics of ultrafast water bridged electron-radical complexes (non independent pairs). From a theoretical and computational point of view, the stochastic modelling of ultra fast radical processes would be upgraded, taking into account the quantum character of nonequilibrium electronic configurations during the *prethermal regime* of very low energy electrons in radiation tracks.

The real-time investigation of interactions between relativistic particle and biomolecular targets opens exciting opportunities for the sensitisation of confined environments (groove of DNA, protein pockets) to ionising radiation on the time scale of molecular motions, i.e. angstrom or sub-angstrom displacements. Potential advances are expected in the next decade, mixing the characteristics of ultra-short micro beams of relativistic particles with those of femtosecond X ray sources. High-time resolved X-ray or electron diffractions of transient states are in progress. The spatio-temporal resolution of transient structures represents some innovating applications of lasers to investigate reaction pathway at the atomic and molecular levels.

Fundamental Aspects of
High Energy Radical Femtochemistry

Theoretical Radiation Chemistry	Time / S	New developments
Energy deposition Stopping power	10^{-18}	Real Time Probing of Initial Events
	10^{-17} 10^{-16}	
Response of matter to energy absorption Dry hole migration (H_2O^+) Physical track structure	10^{-15} 10^{-14}	
	$\mathbf{10^{-13}}$	**Prethermal** **Radical Chemistry**
Mechanism of thermalization and hydration Semi-quantum MD simulations	**5** $\mathbf{10^{-13}}$	Early Molecular Damage
	$\mathbf{10^{-12}}$	
Shape and sizes of track entities Chemical track structure	10^{-11}	
	10^{-10}	
Relationships of partial differential equations to track models in prescribed diffusion and Monte Carlo techniques	10^{-9} 10^{-8}	

Figure 7. Fundamental aspects of high energy radical femtochemistry in aqueous environment. One key point concerns the prethermal processes involved in early radiation-induced molecular damage.

More generally, the quality of short relativistic particle bunches and the combination of low and high-energy femtoradical chemistry would foreshadow advanced developments in spatio-temporal radiation biology and biomedicine [68,69]. This emerging domain concerns notably the highly-selective pro-drug activation from well-defined quantum states of short-lived radicals, the development of real-time nanodosimetry in biologically relevant environment and the investigation of ultra-short irradiation effects on living cells [70].

Acknowledgements
The author thanks several scientists of the *Radiation Biology* MELUSYN Network for fruitful discussions during the manuscript preparation.

References

[1] Turner JE, Magee JL, Wright A, Chatterjee A, Hamm RN and Ritchie RH 1983 *Rad. Res.* **96** 437.
[2] Zemlin F, Schuster R, Beckmann E, Carrascosa JL, Valpuesta JM and Ertle G 1999 *Nature* **399** 51.
[3] Von Sonntag C 2006 *Free-radical-Induced DNA Damage and its Repair* (Heidelberg: Springer).
[4] Horowitz Y 2006 *Microdosimetric Response of Physica and Biological Systems to Low and High Let Radiations: Theory and Appplication to Dosimetry* (Amsterdam: Elsevier).
[5] Elsässer T, Cunrath R, Krämer M and Scholz M *New J. Phys* 2008 **10** 07005.1-07005.17.
[6] Shukla M and Leszczynski J 2008 *Radiation induced molecular phenomena in nucleic acids: a comprehensive theoretical and experimental analysis* (Springer).
[7] Gauduel Y 2008 *Femtochemistry: lasers to investigate ultrafast reactions* in M. Lackner *Lasers in Chemistry* **Vol 2** 861 (Wiley-VCH).

[8] Faure J, Rechatin C, Norlin A, Lifschitz A, Glinec Y and Malka V 2006 *Nature* **444** 737.
[9] Malka V, Faure J, Gauduel Y, Lefebvre E, Rousse A and Ta Phuoc K 2008 *Nature Phys.* **4** 447.
[10] Gauduel Y 1994 Ultrafast electron-proton reactivity in molecular liquids In Simon JD *Ultrafast Dynamics of Chemical Systems* 81 (Kluwer Publisher).
[11] Gauduel Y and Gelabert H 2000 *Chem. Phys.* **256** 333.
[12] Migus A, Gauduel Y, Martin JL and Antonetti A 1987 *Phys. Rev. Lett.* **58** 1159.
[13] Gauduel Y, Pommeret S and Antonetti A 1993 *J. Phys. Chem.* **97** 134.
[14] Henly E and Johnson E 1987 *The Chemistry and Physics of high energy reactions* (University Press).
[15] Laenen R and Roth T 2001 *J. Mol. Structure* **598** 37.
[16] Cramer CJ, Truhlar DG 1994 *Structure and reactivity in aqueous solution*, (ACS Symposium, Washington).
[17] Levine RD, Berstein RB 1987 *Molecular Reaction Dynamics and Chemical Reactivity* (Oxford: University Press).
[18] Moreau M and Turq P 1988 *Chemical reactivity in liquid: fundamental aspects* (Plenum Press)
[19] Gauduel Y and Rossky PJ 1994 *Ultrafast Reaction Dynamics and Solvent Effects* (New York: AIP Press).
[20] Gauduel Y and Borgis D 1996 *Elementary chemical processes in liquids and solutions* Special issue of *J. Chim. Phys.* **93** 1577.
[21] Zewail HA 2000 *J. Phys. Chem. A*, **104** 5660.
[22] Bagchi B 1989 *Ann. Rev. Phys. Chem.* **40** 115.
[23] Allen MP and Tildesley DJ 1987 *Computer simulation of liquids* (Oxford SciencePublications)
[24] Castelman W 2006 Femtochemistry VII *Fundamental ultrafast processes in chemistry, physics and biology* (Elsevier).
[25] Franck F 1985-1989 *Water Science Reviews* (Cambridge University Press).
[26] Laasonen K, Sprik M, Parrinello M and Car R 1993 *J. Chem. Phys.* **99** 9080.
[27] Tuckerman M, Laasonen K, Sprik, M and Parrinello M 1995 *J. Chem. Phys.* **103** 150.
[28] Vuilleumier R and Borgis D 1999 *J. Chem. Phys.* **111** 4251.
[29] Elsaesser T and Bakker HJ 2002 *Ultrafast Hydrogen Bonding Dynamics and Proton Transfer Processes in the Condensed Phase* (Kluwer Publishers).
[30] Bakker HJ, Kropman MF, Omta AW 2004 Ultrafast motion of water molecules near ions In Martin MM and Hynes JT *Femtochemistry and Femtobiology* 149 (Elsevier, Amsterdam).
[31] Agmon N 1998 *J. Phys. Chem.* **102** 192.
[32] Staib A and Borgis D 1996 *J. Chem. Phys.* **104** 9027.
[33] Marcus RA 1993 *Rev. Mod. Phys.* **65** 599 and references therein.
[34] Baxendale JH and Busi F 1982 *The Study of Fast Processes and Transient Species by Electron Pulse Radiolysis* (Dordrecht: Reidel Publishing Company).
[35] Farhataziz and Rodgers M 1987 *Radiation Chemistry* (Weinheim VCH).
[36] Freeman GR 1987 *Kinetics of nonhomogeneous processes* (New York: Wiley).
[37] Goodhead DT 1989 *Int. J. Radiat. Biol.* **56** 623.
[38] Jonah CD and Madhava Rao BSM 2001 *Radiation Chemistry: Present Status and Future Trends* (Amsterdam: Elsevier).
[39] Sabin JR and Brandas E 2007 *Advances in quantum chemistry: theory of the interaction of radiation with biomolecules* (Amsterdam: Elsevier).
[40] Friedland W, Paretzke HG, Ballarini F, Ottolenghi A, Kreth G and Cremer C 2008 *Radiat. Environ. Biophys* **47** 49.
[41] Shukla M and Leszczynski J 2008 *Radiation induced molecular phenomena in nucleic acids* (Springer).
[42] Hart EJ and Anbar M 1970 *The Hydrated Electron* (New York: Wiley).
[43] Jonah CD, Hart EJ and Matheson MS 1973 *J. Phys. Chem.* **77** 1838.
[44] Gopinathan C and Girija G 1983 *Radiat. Phys. Chem.* **21** 209.

Journal of Physics: Conference Series **261** (2011) 012006 doi:10.1088/1742-6596/261/1/012006

[45] Schnitker J and Rossky PJ 1987 *J. Chem. Phys*. ***86*** 3471
[46] Wallqvist A, Martyna G and Berne BJ 1988 *J. Phys. Chem* **92** 1721.
[47] Neria N, Nitzan A, Barnett RN and Landman U 1991 *Phys. Rev. Lett.* **67** 1011.
[48] Tuckerman M, Laasonen K, Sprik M and Parrinello M 1995 *J. Phys. Chem.* **99** 5749.
[49] Yang CY, Wong KF, Skaf MS and Rossky PJ 2001*J. Chem. Phys.* **114** 3598.
[50] Faure J, Glinec Y, Pukhov A, Kiselev S, Gordienko S, Lefebvre E, Rousseau JP, Burgy F and Malka V 2004 *Nature* **431** 541.
[51] Crowell RA, Shkrob IA, Oulianov DA, Korovyanko O, Gosztola DJ, Li Y and Rey de Castro RC 2005 *Nucl. Inst. Met. Phys. Res.* **241** 9.
[52] Oulianov DA, Crowell RA, Gosztola DJ, Shkrob IA, Korovyanko OJ and Rey-de-Castro RC 2007 *J. Appl. Phys***.** **101**, 053102-1-9.
[53] Rechatin C, Faure J, Ben-Ismail A, Lim J, Fitour R, Specka A, Videau H, Tafzi A, Burgy F, and Malka V 2009 *Phys. Rev. Lett.* **102** 164801
[54] Esarey E, Schroeder CB and Leemans WP 2009 *Rev. Mod. Phys.* **81** 1229
[55] Brozek-Pluska B, Gliger D, Hallou A, Malka V and Gauduel Y 2005 *Rad. Phys. Chem.***72** 149.
[56] Gauduel YA, Glinec Y, Rousseau JP, Burgy F and Malka V 2010 *Eur. Phys. J. D* **60** 121.
[57] Gauduel Y, Gelabert H and Guilloud F 2000 *J. Am. Chem. Soc.* **122** 5082.
[58] Gauduel Y, Glinec Y and Malka V 2007 *SPIE Proceedings* **6449** E1.
[59] Wang JCR, Nguyen J and Lu QB 2009 *J. Am. Chem. Soc.* **131** 11320.
[60] Sanche L 2009 *Nature* **461** 358.
[61] Gauduel Y, Berrod S, Migus A, Yamada N and Antonetti A 1988 *Biochemistry* **27** 2509.
[62] Gauduel Y, Sander M and Gelabert H 1998 *J. Phys. Chem. A* **102** 7795.
[63] Gauduel Y, Hallou A and Charles B 2003 *J. Phys. Chem. A* **107** 2011.
[64] Wang CR and Lu QB 2007 *Angew. Chem. Int. Ed.* **46** 6316.
[65] Lu QB 2010 *Mutation. Res. Rev.* **704** 190.
[66] Bergamini, A, Capozzi, M, Ghibelli L, Dini, L, Salanitro, A, Milanese G, Wagner T, Beninati S, Delfina CD, Amici C and Rocchi G 1994 *J. Clin. Invest.* **93** 2251.
[67] Gauduel Y, Glinec Y and Malka V 2008 *J. Phys. CS* **101** 012004.
[68] Malka V, Faure J and Gauduel Y 2010 *Mutation. Res. Rev* **704** 142.
[69] Lacombe S, Sabatier L, Wien F and Gauduel YA 2010 *Cell Death Disease* **1** e4.
[70] Rigaud O, Fortunel NO, Vaigot P, Cadio E, Martin MT, Lundh O, Faure J, Rechatin C, Malka V and Gauduel YA 2010 *Cell Death Disease* **1** e73.

O_2 and glutathione effects on water radiolysis: *a simulation study*

A Colliaux[1,2], B Gervais[3], C Rodriguez-Lafrasse[4] and M Beuve[1,2]
[1] CNRS/IN2P3, UMR 5822, Institut de Physique Nucléaire de Lyon, Université Lyon 1, Villeurbanne, 69622 Lyon, France - [2] LIRIS, UMR CNRS 5205, Université Lyon 1, Villeurbanne, 69622 Lyon, France - [3] Centre de Recherche sur les Ions, les Matériaux et la Photonique - BP 5133 14070 Caen cedex 05, France - [4] Université Lyon 1, Laboratoire de Radiobiologie Cellulaire et Moléculaire, EA 3738, Faculté de Médecine Lyon-Sud, 69921 Oullins, France.

E-mail: m.beuve@ipnl.in2p3.fr

Abstract. We present a MC simulation of water radiolysis when O_2 and glutathione are present in the solution. Our simulation is based on the continuum approximation proposed by Green and co-workers'. We investigate in particular the sensitivity of the yield of $HO_2^{\bullet}+O_2^{-\bullet}$ to the concentration of O_2 and glutathione and to the LET of the ionising radiation. We demonstrate that the production of $HO_2^{\bullet}/O_2^{-\bullet}$ is highly sensitive to these parameters and that the variation of their yield with the O_2 concentration shares remarkable similarities with the O_2 effect reported for biological cell damage efficiency .

1. Introduction

The effects of ionizing radiation in living cells depend on many parameters. Among other factors, oxygen pressure plays a significant role by enhancing cell sensitivity to low-LET radiation. In 1953 Gray *et al* [1] highlighted a correlation between the oxygen concentration and the radioresistance of hypoxic tumors to low-LET irradiation. An issue addressed by numerous studies [2-5] was then to define an "oxygen enhancement ratio" (OER) in order to qualify the cell radiosensitivity as a function of the supplied oxygen pressure (PO_2). The OER was defined as the ratio of the dose required in hypoxic conditions to the dose required at a normal pressure of oxygen, to produce the same biological effect. A compilation by Hall [5] of various experimental data pointed out an increase of the OER with oxygen pressure and a saturation at a pressure of about 30 mmHg. Experiments with neutrons [1] and alpha particles revealed a strong reduction of the oxygen effect with increasing the LET of the radiation. Decreasing the OER and increasing the RBE (Relative Biological Efficiency) were keys to the development of neutron therapy and, more recently, hadrontherapy [6-9]. It is also known that antioxidants could influence the oxygen effect. Indeed the oxygen effect was observed lower in glutathione-deficient cells [10-12]. Glutathione (GSH) is one of the most important antioxidant in the cell. Although the impact of antioxidant and oxidant in the oxygen effect is clear and has been known for a longtime, the mechanisms are still unclear.

The purpose of this work is to quantify by Monte Carlo simulations the influence of an oxygen concentration and an antioxidant concentration in the process of water radiolysis. While the production of hydroxyl radicals by water radiolysis is generally suspected to play a key role in the formation of damage in the important structures of the cells (DNA, membrane…), we will instead focus on the

Journal of Physics: Conference Series **261** (2011) 012007 doi:10.1088/1742-6596/261/1/012007

production of $HO_2^{\bullet}/O_2^{-\bullet}$ radicals. Indeed we shall see that the most important effect of the dissolved oxygen molecules is to scavenge e_{aq}^- and $H^{\bullet}$ radicals leading to a production of $HO_2^{\bullet}/O_2^{-\bullet}$ radicals. This couple of radicals was thought to play a role in the oxygen effect by enhancing the toxicity of ionizing radiations in living cells [13-17]. On a chemical point of view, some pathways are suspected to give it a non-negligible role in damage formation since it can serve as the precursor of an oxidative stress. Indeed, stimulated by iron- or copper-ion catalysis, superoxide anions can rapidly react with hydrogen peroxide producing hydroxyl radicals, through the Haber-Weiss reaction:

$$O_2^{-\bullet} + H_2O_2 \xrightarrow{Fe^{3+}/Fe^{2+}} OH^- + HO^{\bullet} + O_2 \qquad (1)$$

Moreover, it was suggested that a fast reaction of superoxide with nitrogen monoxide (2), which is synthesized by cells, can produce peroxinitrite ions ($ONOO^-$).

$$O_2^{-\bullet} + NO^{\bullet} \longrightarrow ONOO^- \qquad (2)$$

The latter are well known for their deleterious effect on DNA, proteins and lipids. Perhydroxil radical can also efficiently react with nitrogen monoxide, leading to the production of peroxynitrous acid which is also toxic [14]:

$$HO_2^{\bullet} + NO^{\bullet} \longrightarrow ONOOH \qquad (3)$$

Perhydroxyl radical is also known to initiate lipid peroxidation [13]. Alternatively, SuperOxide Dismutase (SOD) enzyme was reported to decrease cell radiosensitivity by inducing the dismutation of superoxide anion into hydrogen peroxide [18, 19] according to:

$$O_2^{-\bullet} + O_2^{-\bullet} \xrightarrow{SOD + 2H^+} O_2 + H_2O_2 \qquad (4)$$

The SOD-related decrease of cell radiosensitivity may therefore be due to the decrease in superoxide-anion concentration. Kuninaka *et al.* [20] reported that a decrease of SOD, which would result in increased quantities of $O_2^{-\bullet}$ in the mitochondria, augmented apoptosis in cell exposed to radiation [20].

Section 2 gives a brief description of the simulation. Section 3 presents the result obtained first in introducing the various concentrations of oxygen, second in combining the effect of an oxygen concentration and an antioxidant (GSH). An effect of LET will be also presented. Section 4 discusses these results and underlines some correlations between the production of the $HO_2^{\bullet}/O_2^{-\bullet}$ radicals and the oxygen effect.

2. Monte Carlo simulation

The simulation aims at predicting the number and the spatial distribution of free radicals and molecules produced in water by a ionizing radiation. The incident particle could be the γ-rays of a Cobalt-60 source or a mono-energetic beam of ions. The γ-rays irradiations were simulated by generating a distribution of electrons for which the energy and the angle were randomly sampled according to the laws of Compton effect, which is the leading process. The irradiations with ions were considered by simulating a segment of ion trajectory. This segment was large enough to get a significant number of chemical species (2.5×10^5 species at 1ps), but small enough to assume that the velocity of the ions could be approximated as constant. To reduce statistical fluctuations, the calculated quantities were averaged over several simulated impacts whatever the incident projectile was.

Journal of Physics: Conference Series **261** (2011) 012007 doi:10.1088/1742-6596/261/1/012007

The principle of our simulation is largely inspired by numerous works [21-24]. A detailed description can be found in [25,26]. The simulation was decomposed in three stages: a physical stage, a physicochemical stage and a chemical stage. To simulate the *physical stage*, the interactions of the ionizing particles with water were described by laws of probability, averaging any effect of water-molecule orientation or local order. The impact of the ionizing particle (Compton electrons or mono-energetic ions) generated a spatial distribution of excited (H_2O^*) or ionic (H_2O^+) water molecules. Ionic species H_2O^{q+} could also be created by multiple ionization. All the ejected electrons were followed in the same way. The electron interactions included not only the processes of ionization and excitation but also the processes of elastic diffusion, electron attachment, and phonon creation. The electrons were followed until their energy fell down to a cut-off energy set to 33 meV (energy of thermalization for a temperature of 300K). Geminate recombination process was also simulated. The simulation output was a spatial distribution of low-energy electrons (thermalized electrons), ionized and excited water molecules.

The *physicochemical stage* describes the medium-relaxation. The thermalized electrons were solvated. The ionized water molecules could either undergo a dissociation process, producing new molecules, or recombine with a thermalized electron to form an excited state. The excited water molecules could also dissociate. When dissociation occurred, the fragments were separated by a specific distance. If multiple-ionization took place, atomic oxygen was produced. The dissociation-channel details and the associated probabilities are available in [26]. In the simulation, all these relaxation channels were sampled by a random process. The outcome of the simulation was a spatial distribution of chemical species. The major species are the hydroxyl radicals, the hydronium ions and the hydrated electrons. These three species mainly issued from single ionization of water molecules according to the channel:

$$\mathbf{H}_2O \xrightarrow{\text{radiation}} \mathbf{H}_2O^+ + e^- \xrightarrow{\text{relaxation}} \mathbf{H}_3O^+ + HO^\bullet + e^-_{aq} \tag{5}$$

The *chemical stage* describes the time evolution of the spatial distribution of the chemical species. Standard chemical models cannot be applied since the spatial distribution can be strongly heterogeneous due in particular to spurs (see figure 1). We therefore developed a Monte Carlo simulation [26], very similar to IRT [21]. This simulation decomposes the whole evolution into a series of short time steps [22-24]. During each time step, the reaction probabilities were calculated for each pair of species as a function of the species diffusion constant, the species charge, the reaction rate, the reaction spin and the reaction radius. A random process selected a part of these reactions according to their reaction probabilities. Since the time step was small, most of the species did not react and simply diffused according to a standard diffusion model. The outcomes of the simulation are the time-dependent radical yields defined as the number of chemical species produced per 100 eV of energy transferred to the sample. Then for a specie I, the yield reads:

$$Y(specie\, i) = 100 \frac{N_i}{\Delta E} \tag{6}$$

where ΔE is energy transferred by the incident particles to the water sample and N_i is the number of species i issuing from this energy transfer.

As the other simulations of heterogeneous chemistry, our simulation *did not include* non-independent-pair dynamics effects. It can therefore not describe the time-dependent yields within the femtosecond regime as measured in recent femtolysis (FEMTOsecond radioLYSIS) experiments [27]. Such dynamics may not affect the yield evolution at times larger than few nanoseconds for low-LET radiations. In the domain of extremely high-LET ions, for which the density of species could be very high, an inclusion of non-independent-pair effects should improve the predictions even at time larger than nanoseconds.

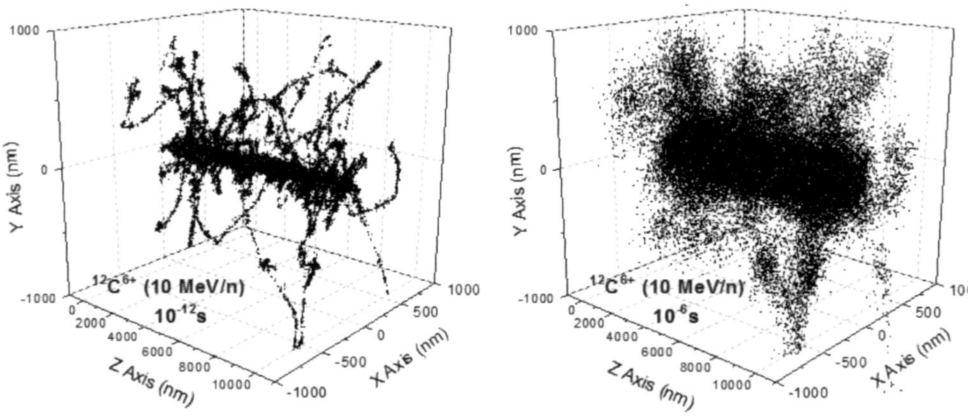

Figure 1. Chemical track-segment of a 10 MeV/n carbon ion simulated in water at 10^{-12} s (left) and 10^{-6} s (right) – Each point represents a chemical species.

We extended the simulation to quantify the effect of a solute on water radiolysis. Since directly introducing the solute to the simulation would dramatically increase the computing time, we applied the approximation introduced by Green *et al* [28,29]. This approximation consists in representing the distribution of solute as a reservoir. Such an approximation has already been considered by other authors [30-32]. The probability for a species to react with the solute reservoir during the period of time t reads:

$$P(t) = 1 - \exp\left(-\alpha\left(x^2 + y^2\right)\right) \tag{7}$$

with:

$$\alpha = \frac{4\pi D\sigma_{eff} C_s N_a}{\beta^2} \tag{8}$$

$$x = \beta\sqrt{t} \tag{9}$$

$$y^2 = \frac{v\sigma}{D}\left(\exp\left(x^2\right)\operatorname{erfc}(x) - 1 + \frac{2}{\sqrt{\pi}}x\right) \tag{10}$$

$$\beta = \frac{v\sigma}{\sigma_{eff}\sqrt{D}} \;;\; \sigma_{eff} = \frac{v\sigma^2}{D + v\sigma} \tag{11}$$

where σ, v and D represent the reaction radius, the reaction velocity and the sum of the diffusion coefficients of both the reactants, respectively. C_s is the concentration of solute and N_a the Avogadro number.

In this work, we have studied the influence of an oxygen concentration on water radiolysis. It is convenient to relate this concentration to the partial pressure of oxygen in air within standard conditions of pressure. We therefore related the concentration and the partial pressure through the Henry laws and considered a Henry constant of 1.3×10^{-3} $Mol.l^{-1}bar^{-1}$, which corresponds to the measured value for O_2 solubility in water at 298°K [33]. Hence, for a partial pressure in air of 160 mmHg, the concentration in oxygen is 0.27 mmol/l. Because we did not simulate the effect of PH and in particular the equilibrium of the $HO_2^\bullet/O_2^{-\bullet}$ acido-basic couple, we could not in principle distinguish between the relative fractions of both these radicals at large time. In this study we also simulated the effect of a concentration of glutathione (GSH). The following reactions were considered:

$$\mathrm{OH}^\bullet + \mathrm{G}SH \longrightarrow H_2O + \mathrm{G}S^\bullet \tag{12}$$

$$\mathrm{H}^\bullet + \mathrm{G}SH \longrightarrow H_2 + \mathrm{G}S^\bullet \tag{13}$$

$$\mathrm{e}_{\mathrm{aq}}^- + GSH \longrightarrow G^\bullet + HS^- \tag{14}$$

These reactions were considered as fully controlled by the diffusion ($v = +\infty$) and their parameters σ (the reaction radius) were deduced from the reaction constant (see table 1). As a first approximation, we did not introduce in the simulation the products of these reactions. We indeed focus in this paper on the first order effects of GSH on water radiolysis. In a second stage, it would be interesting to include the secondary reactions, which is a challenge since these reactions and their constants are not necessary known. Another approximation was to ignore the direct interaction of the radiations with the solute molecules. Excitation and ionization of the GSH and the oxygen molecules should lead to the formation of other radical species. Considering the concentration of these molecules as compared to the number of available water molecules we expect these contributions to the yield of radicals presented in this work to be low.

Table 1. Constant of the reactions with GSH considered in the simulation

Reactions	k (L.mol-1s-1)	Reference
$\mathbf{HO}^\bullet + \mathbf{G}SH \longrightarrow H_2O + \mathbf{G}S^\bullet$	$9.0\mathrm{x}10^9$	34
$\mathrm{H}^\bullet + \mathrm{G}SH \longrightarrow H_2 + \mathrm{G}S^\bullet$	$1.0\mathrm{x}10^{10}$	35
$\mathrm{e}_{\mathrm{aq}}^- + GSH \longrightarrow G^\bullet + HS^-$	$4.7\mathrm{x}10^8$	36

3. Results

3.1. Radiolysis of oxygenated water

We simulated the production of radicals induced by the impacts of low-LET protons in water. At 1 picosecond after the impact, the particle mainly produces $HO^\bullet$, e_{aq}^-, H_3O^+ and in a smaller quantity $H^\bullet$. Because of their close proximity, these radicals may quickly react with each other, creating molecular species such as H_2 and H_2O_2. Figure 2 clearly shows the decrease of $HO^\bullet$ and e_{aq}^-, yields with the time. It also shows results obtained with the same conditions except for the insertion of an oxygen concentration, which corresponds to dissolved oxygen molecules for a normal partial pressure of oxygen in air in standard conditions.

As it is well known, O_2 is an excellent scavenger for solvated electrons and hydrogen radicals. Hence, the most significant changes, with regards to pure water conditions, were marked by a reduction in e_{aq}^- and $H^\bullet$ yields and an increase in $HO_2^\bullet$ and $O_2^{-\bullet}$ yields (see figure 2). The following reactions are involved:

$$H^{\bullet} + O_2 \longrightarrow HO_2^{\bullet} \text{ ; } k = 1.27\ 10^{10}\ L.mol^{-1}.s^{-1} \tag{15}$$

$$e^-_{aq} + O_2 \longrightarrow O_2^{-\bullet} \text{ ; } k = 1.84\ 10^{10}\ L.mol^{-1}.s^{-1} \tag{16}$$

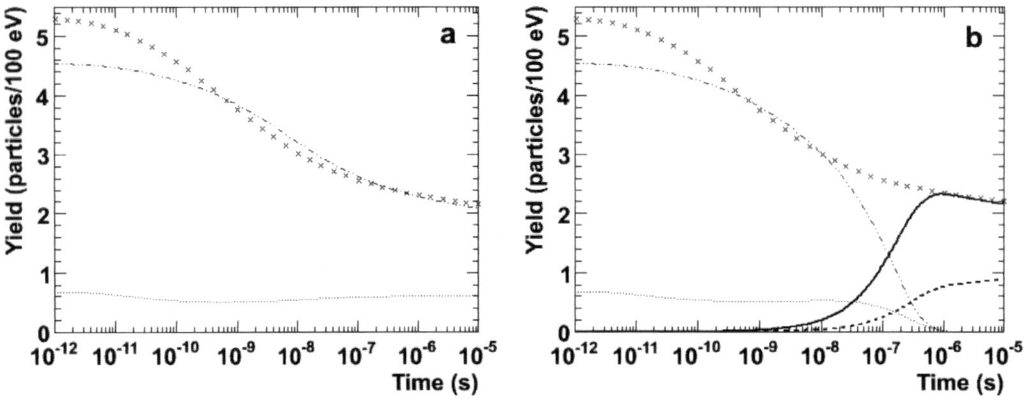

Figure 2. Time-dependant yields of the main species generated by the irradiation with H[65 MeV]of pure water (a) or a partial pressure of oxygen in air of 160 mmHg ($[O_2]$=0.27 mMol.L-1) (b). Species represented are e_{aq}^-(·-·-·-·-) ,$H^{\bullet}$ (·····), $HO^{\bullet}$ (×××), $O_2^{-\bullet}$ (solid line), $HO_2^{\bullet}$ (dashed line).

After 10^{-7} s, the yield of $O_2^{-\bullet}$ in oxygenated water roughly matches that of e_{aq}^- in pure water. $HO_2^{\bullet}$ can also be generated by reaction of $O_2^{-\bullet}$ with hydronium ions H_3O^+:

$$H_3O^+ + O_2^{-\bullet} \longrightarrow HO_2^{\bullet} + H_2O \text{ ; } 4.78\ 10^{10}\ L.mol^{-1}.s^{-1} \tag{17}$$

The production of ($HO_2^{\bullet}$+$O_2^{-\bullet}$) radicals is the major effect of an oxygen concentration in the process of water radiolysis. Figure 3 displays the yield of ($HO_2^{\bullet}$+$O_2^{-\bullet}$) calculated at 10 microseconds for γ-rays as a function of the partial pressure of oxygen in air.

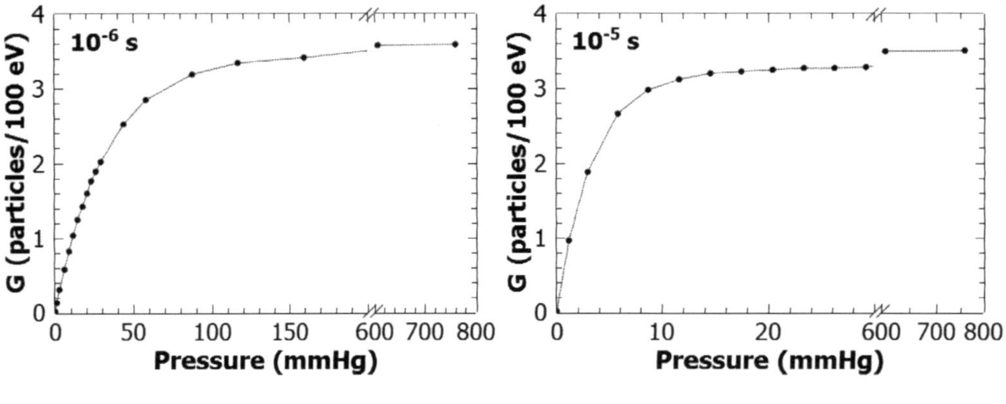

Figure 3. Variation, in function of the partial pressure of oxygen in air, of the yield in ($HO_2^{\bullet}$+$O_2^{-\bullet}$) produced at 1 μs (left) and 10 μs (right) by the γ-rays of a Cobalt-60 source (LET~0.3 keV/μm).

One can observe a strong increase in the yield for pressures lower than about 10 to 30 mmHg and a quasi-saturation for pressures larger to this threshold. of the e_{aq}- and $H^{\bullet}$ radicals and therefore increased the yield of $(HO_2^{\bullet}+O_2^{-\bullet})$ at 10 microseconds. The saturation pressure corresponded to the minimal pressure value at which most of the e_{aq}- and H Below the pressure of saturation, a significant quantity of e_{aq}^- and $H^{\bullet}$ radicals were not scavenged by the oxygen molecules before 10 microseconds. Increasing the pressure fastened the consummation $^{\bullet}$ radicals were scavenged by the oxygen molecules at 10 microseconds. Figure 3 shows also the same yield calculated at 1 microsecond. For both times, the curves are qualitatively similar. However, the saturation pressure was larger for shorter times: as expected, a larger pressure is required to scavenge most of the e_{aq}^- and $H^{\bullet}$ radicals a shorter time after the particle impact.

While the time at which the yield is calculated significantly affects the evolution with the pressure, fixing it appears somewhat arbitrary. In particular, in a biological system, there is no reason to consider a specific observation time and the production of $(HO_2^{\bullet}+O_2^{-\bullet})$ has to be considered over a larger time. However this production is expected to be limited by the lifetime of the e_{aq}^- and $H^{\bullet}$ radicals in the cells.

3.2. Radiolysis of oxygenated water with a concentration of GSH

Anti-oxidants have a major protective role in cells by scavenging the free radicals. The level of antioxidants controls therefore the lifetime of radicals. We expect therefore the production of $(HO_2^{\bullet}+O_2^{-\bullet})$ to be modified by a concentration of antioxidants. We therefore inserted a concentration of GSH into our simulation system. We were particularly interested in GSH because it is known to be one of the most important antioxidants of the cells. Moreover it was reported that the oxygen effect was sensitive to the level of GSH in the cells [10-12]. GSH can be found both in the cell nucleus and cell cytoplasm [37]. We first calculated the time-dependent yields of the main radicals for a concentration of GSH of 1 mmol.L^{-1} and 10 mmol.L^{-1}, which correspond to typical concentrations in normal cells [38]. The calculations were performed for a Cobalt-60 source. The main effect is the scavenging of $HO^{\bullet}$ radicals. However as can be seen in figure 4, the scavenging of e_{aq}^- and $H^{\bullet}$ radicals is also quite significant at such concentrations and the lifetime of e_{aq}^- and $H^{\bullet}$ radicals are reduced by GSH to values lower than 10 microseconds. The lifetime of e_{aq}^- radicals is larger than the lifetime of $H^{\bullet}$ radicals due to a lower reaction constant (see table 1). As a consequence of this scavenging, the production of $(HO_2^{\bullet}+O_2^{-\bullet})$ in oxygenated water by reactions with e_{aq}^- and $H^{\bullet}$ radicals is achieved before 10 microseconds at such GSH concentrations.

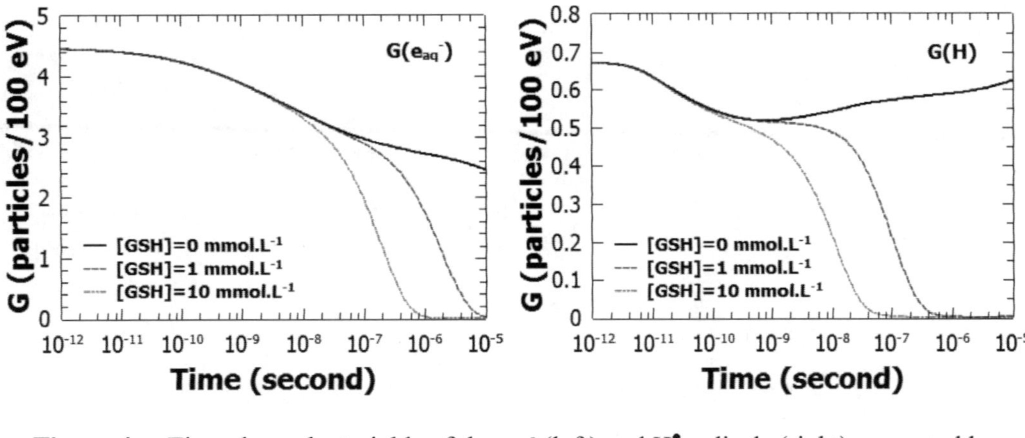

Figure 4. – Time-dependant yields of the e_{aq}^- (left) and $H^{\bullet}$ radicals (right) generated by the γ-rays of a Cobalt-60 source (LET~0.3 keV/μm) in water with a concentration in GSH of 0,1 and 10 mmol.L-1.

Figure 5 combines the effect of both the dissolved oxygen and the GSH concentration. It represents the yield of ($HO_2^{\bullet}+O_2^{-\bullet}$) as a function of the partial pressure of oxygen in air. The calculations were performed for a Cobalt-60 source with GSH concentration of 1 mmol.L^{-1} and at a time large enough as compared to the lifetime of e_{aq}^- and $H^{\bullet}$ radicals.

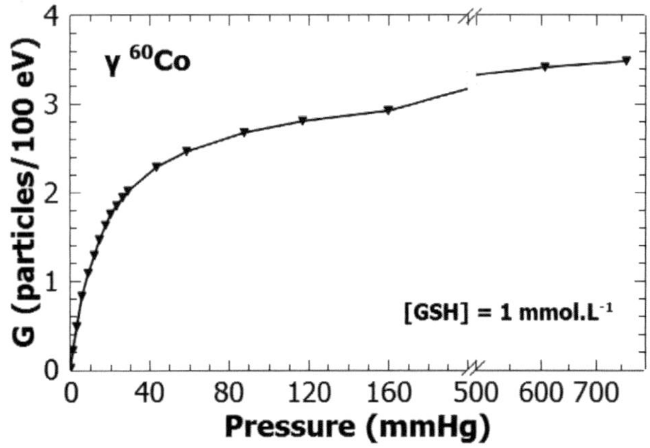

Figure 5. Variation with the partial pressure of oxygen in air of the yields of $HO_2^{\bullet}$ + $O_2^{-\bullet}$ produced at 10 μs by the γ-rays of a Cobalt-60 source (LET~0.3 keV/μm) in water with a concentration in GSH of 1 mmol.L-1.

As can be seen on figure 5, the yield of ($HO_2^{\bullet}+O_2^{-\bullet}$) increased with the pressure and saturated. These features have to be attributed to the competition between the scavenging by the oxygen molecules and the scavenging by the GSH molecules. Below the pressure of saturation, the scavenging by GSH is dominant. Increasing the pressure of oxygen fastened the scavenging by oxygen and increased the production of ($HO_2^{\bullet}+O_2^{-\bullet}$). The latter production saturated as soon as the scavenging by oxygen got dominant and could capture most of the available e_{aq}^- and $H^{\bullet}$ radicals.

A last parameter we considered in this study was the effect of the LET. Figure 6 shows the yield of ($HO_2^{\bullet}+O_2^{-\bullet}$) generated by the impacts of proton and carbon ions in water oxygenated by a partial pressure of 160 mmHg in air and containing a concentration in GSH of 1 mmol.L^{-1}.

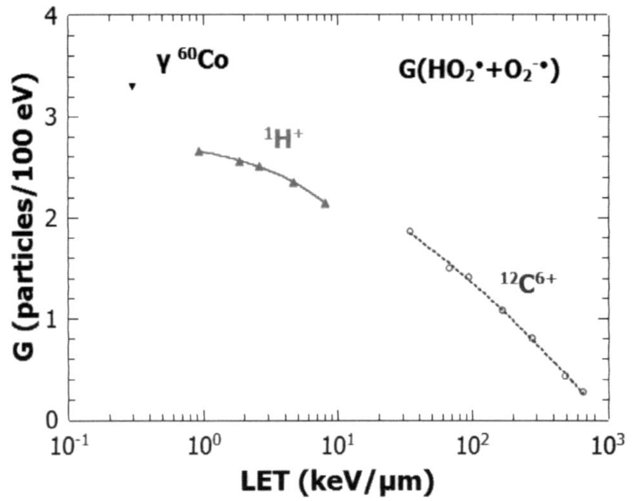

Figure 6. Yield of ($HO_2^{\bullet}+O_2^{-\bullet}$) expressed as a function of the LET for the γ-rays of a Cobalt-60 source (▼), protons (–▲–) and carbon ions (---○---). The calculations were performed at 10 microseconds for a partial pressure of oxygen of 160 mmHg and a concentration in GSH of 1 mmol.L^{-1}.

We can clearly observe that increasing the LET reduced the production of ($HO_2^{\bullet}+O_2^{-\bullet}$). The yield fell down to almost zero for the highest LET carbon projectile. This strong reduction arose from the large concentrations of radicals produced in the track (see figure 1). Due to their close proximity, the radicals interacted with each other before any significant scavenging by the oxygen molecules. The production of ($HO_2^\bullet$+$O_2^{-\bullet}$) thus became a minor process for high LET.

4. Concluding remarks and perspective

As summarized in the introduction section, superoxide anion ($O_2^{-\bullet}$) and its protonated form, the hydroperoxyl radical ($HO_2^\bullet$) were suspected to play a role in the oxygen effect by enhancing the toxicity of ionizing radiation in living cells [13-17]. The related chemical and biological arguments were a motivation for us to study the parameters that may influence the production of $HO_2^\bullet/O_2^{-\bullet}$. We modified our simulation of the water radiolysis by ionizing particle to insert a concentration of dissolved oxygen and a concentration of antioxidant. As antioxidant we considered glutathione (GSH) since it is one of the most important antioxidants in living cells. Moreover it was reported that the oxygen effect was sensitive to the level of GSH in the cells [10-12]. We have shown that the production of $HO_2^\bullet/O_2^{-\bullet}$ was modified by the presence of GSH. At concentrations ranging from 1 to 10 mmol.L^{-1} and corresponding to typical concentrations in GSH for normal cells [38], the scavenging of the e_{aq}^- and $H^\bullet$ radicals was quite significant. This reduction in the yield of e_{aq}^- /$H^\bullet$ strongly affected the production of $HO_2^\bullet/O_2^{-\bullet}$, for which the production mainly issued from the reaction of the e_{aq}^- and $H^\bullet$ radicals with the oxygen molecules. The competition between the scavenging of e_{aq}^- and $H^\bullet$ radicals by the oxygen molecules and by the GSH concentrations led to particular features when plotting the yield of ($HO_2^\bullet$+$O_2^{-\bullet}$) as a function of the dissolved oxygen concentration (or equivalently the partial pressure of oxygen in air). At low oxygen pressure, the scavenging by GSH dominated. Increasing the pressure of oxygen fastened the scavenging of the e_{aq}^- and $H^\bullet$ radicals by oxygen and led to an increase of the production of $HO_2^\bullet/O_2^{-\bullet}$. We observed however that this increase in the $HO_2^\bullet/O_2^{-\bullet}$ production saturated as soon as the scavenging by oxygen became predominant. For a concentration of 1 mmol.L^{-1} in GSH, the pressure of saturation amounted to about 20 to 30 mmHg.

From a quantitative point of view, the yield of ($HO_2^\bullet$+$O_2^{-\bullet}$) was found as large as 0.3 μmol.L^{-1} per Gray for a normal pressure of oxygen (160 mmHg) and with a concentration of 1 mmol.L^{-1} in GSH. The production is therefore several orders of magnitude larger than the standard rate in non-irradiated cells for which the concentration at equilibrium varies between 0.01 nmol.L^{-1} to 0.1 nmol.L^{-1} [14, 39]. We also showed that the production of $HO_2^\bullet/O_2^{-\bullet}$ was sensitive to the type of radiation. Increasing the LET of the particle could strongly decrease their production. This reduction was due to the fast reactions between the radicals, which are produced in close proximity for high LET radiation.

Although the comparison between a living cell and a simple system as the one we simulated is limited, we attempt a parallel in the evolution of the production of $HO_2^\bullet/O_2^{-\bullet}$ and the oxygen effect with the studied parameters. A common point is the evolution with the pressure of oxygen. Both increase with the oxygen pressure and saturate. The saturation pressure seems lower for the oxygen effect. It amounts to 20 mmHg for the production of $HO_2^\bullet/O_2^{-\bullet}$ at a concentration of 1 mmol.L-1 in GSH and to 5 mmHg for the oxygen effect in fibroblast cells Solen *et al.* [11]. For the latter the concentration in GSH was however not estimated.

Another common feature of the oxygen effect and the production of $HO_2^\bullet/O_2^{-\bullet}$ is the decrease with the LET and the disappearance at very high-LET. The decrease in the production of $HO_2^\bullet/O_2^{-\bullet}$ is related to the large concentration of energy deposited along the ion trajectory, which leads to efficient reactions between the free radicals and therefore to an inefficient scavenging by the oxygen molecules.

Caution must be exercised in comparing these results because of the complexity of living cells. Many other elements in a cell could indeed modify the production of $HO_2^\bullet/O_2^{-\bullet}$ either by reducing it or by increasing it. Moreover we did not insert in our system the other forms of glutathione (such as GSSG), nor the products of reactions ($G^\bullet$, $HS^\bullet$, $GS^\bullet$). The latter could also react with oxygen and contribute to the toxicity of the radiations in oxygenated conditions. Moreover, glutathione contributes to cell response to radiations by reducing the damage capacity of $HO^\bullet$ radicals. The direct action of the radiations to the cells has also to be considered.

Finally by including the effects of two natural solutes on water radiolysis, this work constitutes one step toward the complexity of radiobiology. Clearly other solutes should be introduced in the simulation. To this aim a systematic determination of the reaction parameters for the most important solutes will be required. This includes the reactions of the solute with the water radicals but also the cascades of reactions that may follow. Researches, both experimental and theoretical, are still necessary to understand the origin(s) of the oxygen effects. The interest is not only important on a fundamental point of view, but also for applications in Oncology since it could contribute to an improved treatment of hypoxic tumors [40-42].

Acknowledgment
This work was supported by the CPER hadrontherapy of the Lyon/Rhône-Alpes, by the ANR RADICO and by the French National Institute of Cancer (INCa). This support is gratefully acknowledged.

References

[1] Gray L H, Conger A D, Ebert M, Hornsey S and Scott O C A 1953 *Br. J. Radiol.* **26** 638.
[2] Alper T and Howard-Flanders P 1956 *Nature* **178** 978.
[3] Alper T 1956 *Radiat. Res.* **5** 573.
[4] Blakely E A, Ngo F Q, Curtis S B and Tobias C A 1984 *Adv. Radiat. Biol.* **11** 295-389
[5] Hall E J 2000 *Radiobiology for the radiobiologist. 5th ed* Lippincott Williams & Wilkins (Philadelphia PA)
[6] Hirao Y, Ogawa H, Yamada S, Sato Y, Yamada T, Sato K, Itano A, Kanazawa M, Noda K and Matsumoto S 1992 *Nucl. Phys. A* **538** 541.
[7] Haberer Th, Debus J, Eickhoff H, Jäkel O, Schulz-Ertner D and Weber U 2004 *Radiother. Oncol.* **73** (suppl 2) S186.
[8] Amaldi U, 2004 *Radiother. Oncol.* **73** (suppl 2) S191.
[9] Bajard M, De Conto J-M, Remillieux J and ETOILE *Radiother. Oncol.* **73** (suppl 2) S211.
[10] Hedgren M, Larsson A, Nilsson K, Révész L and Scott O C 1980 *Int. J. Radiat. Biol.* **37** 299.
[11] Solen G, Edgren M, Scott O C A and Révész L 1987 *Int. J. Radiat. Biol.* **51** 39.
[12] Solen G, Edgren M, Scott O C A and Révész L 1989 *Int. J. Radiat. Biol.* **55** 201.
[13] Aikens J and Dix T A 1991 *J. Biol. Chem.* **23** 15091.
[14] Goldstein A and Czapski G 1995 *Free. Rad. Biol. Med.* **19**(4) 505.
[15] De Grey *ADN J.* 2002 **21**(4) 251.
[16] Mikkelsen R B and Wardman P 2003 *Oncogene.* **22** 5734.
[17] Delattre J, Beaudeux J-L et Bonnefont-Rousselot D 2005 *Radicaux libres et stress oxydant. Aspects biologiques et pathologiques* (Éditions Lavoisier Paris)
[18] Josson S, Xu Y, Fang F, Dhar S K, St Clair D K and St Clair W H 2006 *Oncogene* **25** 1554.
[19] Borrelli A, Schiattarella A, Mancini R, Morrica B, Cerciello B, Mormile M, d'Alesio V, Bottalico L, Morelli F, D'Armiento M, D'Armiento F P and Mancini A 2009 *Free Radic. Biol. Med.* **46** 110.
[20] Kuninaka S, Ichinose Y, Koja K and Toh Y 2000 *Br. J. Cancer.* **83**(7) 928.
[21] Green N J B, Pilling M J, Pimblott S M and Clifford P 1990 *J. Phys. Chem.* **94**, 251.
[22] Pimblott S M and La Verne J A 1997 *J. Phys. Chem A* **101** 5828.
[23] Cobut V, Frongillo Y, Patau J P, Goulet T, Fraser M-J and Jay-Gerin J-P 1998 *Radiat Phys. Chem.* **51** 229.
[24] Frongillo Y, Goulet T, Fraser M-J, Cobut V, Patau J P and Jay-Gerin J-P 1998 *Radiat. Phys. Chem.* **51** 245.
[25] Gervais B, Beuve M, Olivera G H, Galassi M E and Rivalora R D 2005 *Chem. Phys. Lett.* **410** 330.
[26] Gervais B, Beuve M, Olivera G H and Galassi M E 2006 *Radiat. Phys. Chem.* **75** 493.
[27] Gauduel Y A, Glinec Y, Rousseau J P, Burgy F and Malka V 2010 *Eur. Phys. J. D* **60**, 121.
[28] Green N J B 1984 *Chem. Phys. Lett.* **107** 485.
[29] Pimblott S M, Pilling M J and Green N J B 1991 *Radiat. Phys. Chem.* **37** 377.

[30] Tomita H, Kai M and Kusama T 1997 *Environ. Biophys.* **36** 105.
[31] Watanabe R and Saito K 2001 *Radiat. Phys. Chem.* **62** 217.
[32] Cobut V, Corbel C and Patau J-P 2005 *Radiat. Phys. Chem.* **72** 207.
[33] Wilhelm E, Battino R and Wilcock R J 1977 *Chem. Rev.* **77** (2) 219.
[34] Babbs C F and Steiner M G 1990 *Free Radic. Biol. Med.* **8** 471.
[35] Zhao R, Lind J, Merényi G and Eriksen T E 1997 *J. Chem. Soc., Perkin Trans. 2* 569-74
[36] Hoffman M Z and Hayon E 1973 *J. Phys. Chem.* **8** 990.
[37] Voehringer D W, McConkey D J, McDonnell T J, Brisbay S and Meyn R E 1998 *Proc. Natl. Acad. Sci. USA* **95** 2956.
[38] Schafer F Q and Buettner G R 2001 *Free Radic. Biol. Med.* **30** 1191.
[39] Salvador A, Sousa J and Pinto R E 2001 *Free. Radic. Biol. Med.* **31** 1208.
[40] Vaupel P and Harrison L 2004 *Oncologist.* **9** (suppl 5) 4.
[41] Carlson D J, Stewart R D and Semenenko V A 2006 *Med. Phys.* **33** 3105.
[42] Sovik A, Malinen E, Bruland O S, Bentzen S M and Olsen D R 2007 *Phys. Med. Biol.* **52** 499.

Journal of Physics: Conference Series **261** (2011) 012008 doi:10.1088/1742-6596/261/1/012008

NEXAFS : a unique tool to follow the photochemistry of small organic molecules in condensed water

Ph Parent[1], C Laffon[1], F Bournel[1], J Lasne[1] and S Lacombe[2]
[1]Laboratoire de Chimie–Physique, Matière et Rayonnement, Université Pierre et Marie Curie (UPMC–Univ Paris 06) and CNRS (UMR 7614), 11 rue Pierre et Marie Curie, 75231 Paris Cedex 05, France - [2]Institut des Sciences Moléculaires d'Orsay, ISMO (FRE 3363), 91405 Orsay Cedex, France and Université Paris Sud 11, CNRS-Bât 351, 91405 Orsay Cedex, France.

E-mail: philippe.parent@upmc.fr

Abstract. Soft X–ray induced chemistry of simple organic molecules as carbon monoxide (CO), methanol (CH_3OH) and glycine (NH_3CH_2COOH) in water and nitric acid hydrate has been investigated with NEXAFS spectroscopy. In the pure species, extremely high survival rates are observed, a consequence of the back reactions allowed in the condensed phase. When mixed with water, the survival rates are considerably reduced by reaction with the hydroxyl radical (OH). The formation of CO_2 is also enhanced at the expense of CO, the main byproduct in the photolysis of the pure species. Finally, it is shown that water plays no role in the destruction of the amino acid.

1. Introduction

Radiation damages of organic molecules in water are of fundamental importance in radiobiology. They are caused by the direct interaction with the radiation and by chemical attack of the target molecule by the reactive oxygenated species (ROS) released by water, especially OH [1,2]. We have recently used the X–ray absorption spectroscopy (NEXAFS) for studying these processes at a molecular scale, starting with pure water ice [3], water doped with O_2 [4], and ammonia ice [5]. The advantage of NEXAFS is to display all the byproducts created after irradiation, regardless of their chemical form (*i.e.*, atoms, molecules, radicals), with spectral intensities only related to their concentrations, which allows identifying the dominant reaction channels. In the present work, we study further simple (CO) and more complex molecules condensed in water (and nitric acid hydrate), like methanol (CH_3OH) and glycine (NH_3CH_2COOH), the simplest amino acid. The comparisons between the irradiation of the pure species and the mixes with H_2O reveal the interplay between direct photolysis and indirect chemistry induced by the ROS in H_2O.

2. Experimental

The experiments on methanol were performed at MAX–Lab (MAXII, Lund, Sweden) on the I311 undulator beamline equipped with a SX–700 monochromator. The experiments on CO:H_2O mixtures were performed at LURE (SuperACO, Orsay, France) on the SA22 bending magnet beamline equipped with a plane grating spherical mirror monochromator. The experiments on glycine were recorded at the Synchrotron SOLEIL (Saint Aubin, France) on the TEMPO beamline equipped with a plane grating spherical mirror monochromator. The C and O K edges were recalibrated after the experiments in order to match the π^* absorption features of CO at 287.3 eV for the C K edge and 534.1 eV for the O K edge. The sample preparation and the NEXAFS measurements have been performed in a similar manner in the three experiments, except the irradiations. They were done at LURE using the unmonochromatized white beam delivered at the zero–order of the monochromator (the bandwidth is then 1–1000 eV), and with 150 and 530 eV at MAX–II and SOLEIL, respectively. The absorbed doses were estimated from the photocurrent measured on the sample (LURE) and from a calibrated photodiode (MAX–II, SOLEIL). All the experiments were achieved under ultrahigh vacuum (base pressure 1×10^{-10} mbar) by condensing pure or mixed vapors on a metallic substrate cooled by a liquid helium cryostat. For CH_3OH and H_2O, the gases were introduced from the vapor pressure of the liquids (after purification by several freeze–pump–thaw cycles) using separated circuitries in case of mixed ices. Ultrapure CO (99.997 %) gas was provided by Air Liquide. The nitric acid hydrate ice film was obtained by directly dosing the vapor from the azeotrope mixture of 68 wt % HNO_3 in H_2O (about $HNO_3.1.65H_2O$ of molar composition). The partial pressures were calibrated both in the primary dosing circuitry with Pirani gauges and by mass spectrometry into the UHV vessel. The sticking coefficient being 1 for all the compounds at the deposition temperatures, these pressures directly providethe composition of the films. Depending on the case, the film thickness was typically 120 and 200 monolayers (ML), controlled by the exposure times. This is thick enough to ensure that the probed depth (typically 50–100 Å) is not perturbed by the substrate. The glycine/H_2O mixtures were prepared from the vapor pressure of the pure components and/or by mixing the appropriated gas phase, adjusting the water vapor pressure to obtain the desired glycine dilution, followed by the co–deposition on a cold substrate maintained at 30 K. Vaporization of glycine was achieved using a Knudsen cell held at 413 K under UHV. The deposition rates were monitored to deposit about 600 ML.

3. Results and discussion

3.1. CO/H_2O

Figure 1a presents the normalized C K–edge NEXAFS spectra of three CO:H_2O mixtures: 100:0 (pure CO), 66:34 and 24:76, deposited and irradiated at 20 K. The peak at 287.3 eV is the C1s → π^*(CO) transition [6]. It decreases after irradiation as a result of the CO destruction. One main feature appears at 290.7 eV, which is the C1s → π^* transition of CO_2 [6]. This indicates that CO is mainly converted into CO_2 [7]. On the 66:34 and 24:76 films, the π^*(CO) is smaller and the $\pi^*(CO_2)$ larger. Thus, mixing CO with water amplifies the destruction and enhances the formation of CO_2. In figure 1b are displayed the remaining part (in %) of the initial π^*(CO) intensity as a function of the absorbed dose, which rates the survival of CO at the various dilutions. The survival rates are modelled accounting for the balance between the destruction reaction (1) and the back reactions (2):

$$CO + (h\nu, e^-) \rightarrow C + O \quad (1)$$

$$C + O \rightarrow CO \quad (2)$$

where ($h\nu$, e^-) indicates that the CO may decompose either through direct absorption of the incident photons or by interaction with secondary electrons emitted on the track of the photons. The CO concentration can be written:

$$dc/dt=\{-\sigma_d.c+ \sigma_f(1-c)\}.\Phi \quad (3)$$

where c=[CO], σ_d and σ_f are the reaction yields of reactions 1 and 2, respectively; c_0 is the initial CO concentration, Φ the photon flux and t the exposure time (Φ.t is proportional to the absorbed dose). By integration of equation 3 we get:

$$c = c_0 \{(\sigma_f/(\sigma_d-\sigma_f))+(1-(\sigma_f/(\sigma_d-\sigma_f))\exp(-(\sigma_d-\sigma_f).\Phi.t\} \quad (4)$$

Journal of Physics: Conference Series **261** (2011) 012008 doi:10.1088/1742-6596/261/1/012008

Equation 4 is used to fit the survival rate presented in figure 1b; it is in excellent agreement with the experiments. Concerning the production of CO_2 in pure CO, it can be formed after reaction 1 by addition of O to a neighboring CO molecule [8]:

$$CO + O^* \rightarrow CO_2 \tag{5}$$

And also by the Boudouard reaction [7–10]:

$$CO + (h\nu, e^-) \rightarrow CO^* \tag{6}$$

$$CO + CO^* \rightarrow CO_2 + C \tag{7}$$

When mixed with water, CO_2 can also be formed by the addition of OH to CO [8]:

$$CO + OH \rightarrow CO_2 + H \tag{8}$$

This reaction strongly reduces the CO concentration and hence, lowers its survival rate. It also increases the CO_2 concentration. Figure 1c presents the ratio of the $\pi^*(CO_2)$ over the $\pi^*(CO)$ areas, a quantity proportional to the CO_2/CO ratio. In the 24:76 film, the formation of CO_2 is 8 times more *efficient than in pure CO.*

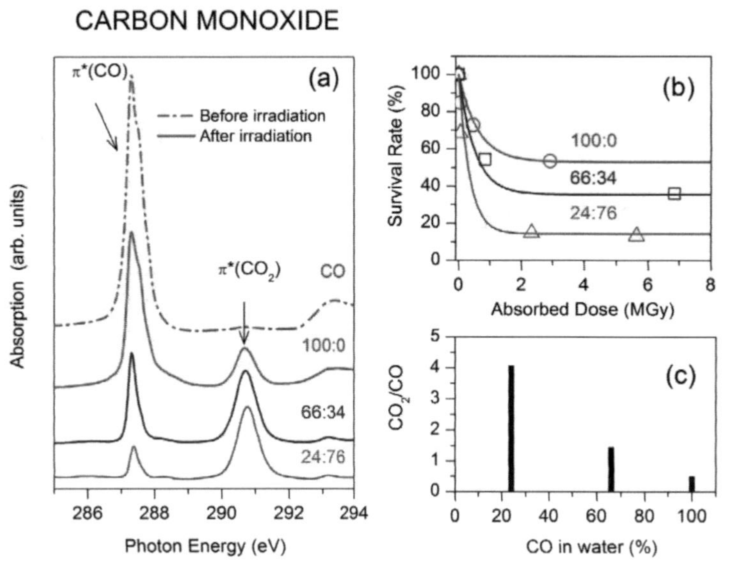

Figure 1. (a) C K-edge NEXAFS spectra of $CO:H_2O$ mixtures: 100:0, 66:24 and 24:76, irradiated at 20 K (dotted line: before irradiation, full line : after irradiation. (b) Survival rate of CO in as function of the absorbed dose; experimental data and fit using Eq. (4) (full line). (c) The CO_2/CO ratios as function of the CO concentration in water ice, calculated from the intensity ratios of the π^* transition of the two species

3.2. CH_3OH/H_2O and HNO_3.

Figure 2a presents the C K–edge NEXAFS spectra of pure methanol (100:0), methanol in water (50:50, 16:84), and 34% of methanol into 66 % of $HNO_3.1.65H_2O$ (34:66 NitHyd). The films have been deposited at 60 K and irradiated at 20 K with a photon energy of 150 eV. Methanol before irradiation (dotted lines) shows two peaks labeled σ^* at 288.5 eV and π^* at 289.7 eV. They decrease as methanol is destroyed. Two main peaks appear then at 287.3 eV and 290.7 eV, which are the π^* transition in CO and CO_2, respectively. This shows that methanol is mainly converted into CO and

CO_2. As observed previously for CO, the $\pi^*(CO)$ transition becomes smaller and the $\pi^*(CO_2)$ larger as the water concentration increases.The most probable photolysis reaction is [11]:

$$CH_3OH \rightarrow CH_3O + H \quad (9)$$

Then, methanol can recombine:

$$CH_3O + H \rightarrow CH_3OH \quad (10)$$

This back reaction easily occurs, since it is barrierless [12]. The survival rates of methanol in the various mixes have been calculated by the relative decrease of the spectral intensity at 288.5 eV, where only methanol contributes. They are plotted on figure 2b as function of the absorbed dose. These data are also fitted with equation 4 (with $c=[CH_3OH]$), accounting for a balance between reactions 9 and 10. In pure methanol, about 50 % of the initial methanol concentration has survived to the irradiation. This rate drops when mixed with water (30 % in the 16:84 film), and strongly decrease to 5 % of survival in presence of HNO_3.

In pure methanol, the fact that CO is the main byproduct indicates that irradiation essentially causes dehydrogenation, through a reaction scheme that has been investigated in detail by Bennett et al. [12], schematically condensed in reaction 11:

$$CH_3OH + 4(h\nu, e^-) \rightarrow CO + 4H \quad (11)$$

In this reaction, low energy electrons are very efficient [13]. CO formed by reaction 11 is then transformed in CO_2, as discussed in the previous section (reactions 5–7).

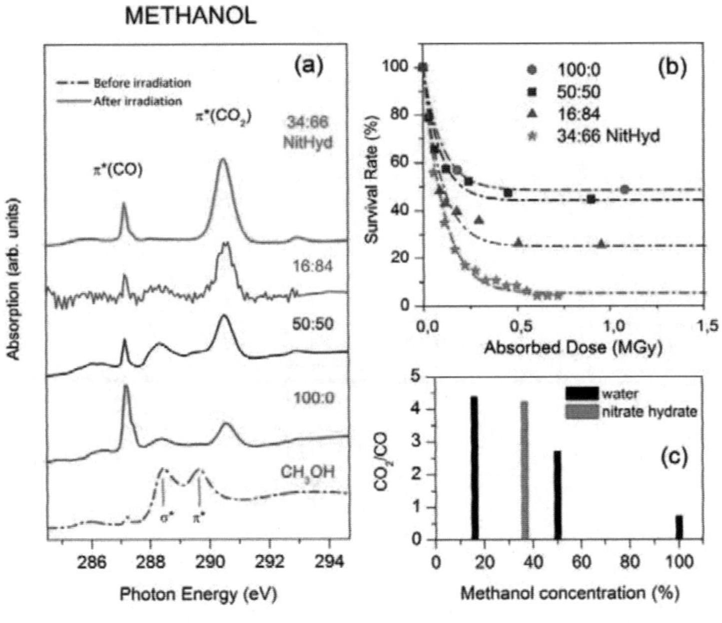

Figure 2. C K-edge NEXAFS spectra of pure methanol (100:0), methanol in water (50:50, 16:84), and 34% of methanol into 66 % of $HNO_3.1.65H_2O$ (34:66 NitHyd). The films have been deposited at 60 K and irradiated at 20 K and 150 eV. (b) Survival rates of methanol in the four samples, as function of the absorbed dose; experimental data and fit using Eq. (4). (c) CO_2/CO ratios as function of the methanol concentration, calculated from the intensity ratios of the π^* transition of the two species.

In water, the methanol photolysis has been reviewed in ref. [14]. In brief, most of the reactions occur with the OH radical to form intermediate dehydrogenated methanol species H_xCO and end in CO. This is schematically condensed in reaction 12:

$$CH_3OH + 4OH \rightarrow CO + 4H_2O \tag{12}$$

At the end of reaction 12, CO will further react with OH (reaction 8), as observed for CO in ice, which explains the enhanced formation of CO_2 in water compared to pure methanol. This is illustrated in figure 2c, which presents the CO_2/CO ratio for the various films. We see that for methanol diluted in water at 16 %, the formation of CO_2 is 6 times more efficient than in pure methanol.

In the nitric acid hydrate sample, the solvated NO_3^- ion is readily photo–decomposed into NO_2^-+O and NO_2+OH [15,16], which further increases the OH and the O concentrations, then the destruction of methanol by reaction (12) (and a similar dehydrogenation reaction with O). It ends in CO, which, again, will react with O and OH to form CO_2, increasing the CO_2/CO ratio to 4.2 for the 34:66 methanol/nitric acid hydrate mixture, a ratio obtained for methanol in pure water ice at higher dilution (16:84) (figure 2c).

3.3. Glycine/H_2O

Soft X-ray irradiation of glycine has been already studied using NEXAFS spectroscopy, but not at low temperature in ice [17,18,19].Figure 3a presents the O K–edge NEXAFS spectra of pure glycine (top) and of 15% of glycine in H_2O ice (15:85, bottom) before and after irradiation at 530 eV at 20 K. The fingerprint of glycine is the narrow peak at 532.5 eV, labeled Gly COO^-, which is the O1s→π*(COO^-) excitation in the zwitterionic glycine [20]. In water, the broad band centered at 540 eV is the conduction band of ice, and the shoulder at 535 eV is the O1s→σ*(O–H) ($3a_1$) excitation in H_2O [21].

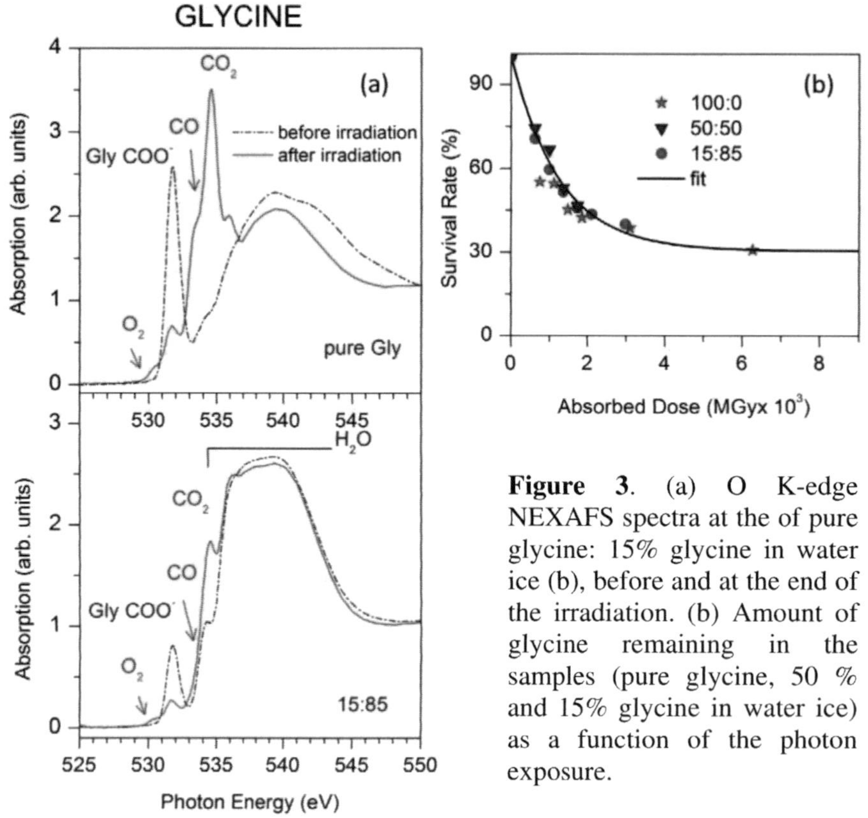

Figure 3. (a) O K-edge NEXAFS spectra at the of pure glycine: 15% glycine in water ice (b), before and at the end of the irradiation. (b) Amount of glycine remaining in the samples (pure glycine, 50 % and 15% glycine in water ice) as a function of the photon exposure.

Under irradiation, the $\pi^*(COO^-)$ intensity decreases in the two samples, resulting from the glycine destruction. At the same time, two main peaks grow at 534.3 eV and 535.4 eV, characteristic of CO and CO_2, respectively, a result of the removal of the carbonyl group [19]. The CO_2/CO peak intensities ratio is 3 times higher for glycine in ice than in pure glycine (not presented), an indication that, again, water favors the formation of CO_2 at the expense of CO (reaction (8)). The rate of destruction of glycine can be derived from the decrease of the $\pi^*(COO^-)$ intensity (figure 3b). This time, the survival rate of glycine is fairly the same in pure glycine and in glycine in ice. In all the cases, it tends to a limit of about 30% of the initial glycine concentration. Quantum calculations have shown that the reaction of isolated glycine with OH is possible, under the restricted conditions that the OH radical approaches with a specific reaction coordinate that facilitate the reaction [22]. Our results show that for glycine in ice, such reaction geometry is unlikely. Thus, contrary to what is observed for CO and methanol in H_2O, water plays *no role* in the destruction of glycine.

4. Conclusion

This study shows that the condensed phase allows back–reactions that enhance the survival rates, a strong limit to the photolysis efficiency. In the pure CO film, CO is transformed mainly into CO_2 by CO+O. In presence of water, the CO+OH reaction enhances the CO destruction and increases the CO_2/CO ratio. CH_3OH is also mainly transformed into CO by successive dehydrogenations, and then into CO_2 by CO+O. In presence of water, methanol is further transformed into CO by an additional sequence of dehydrogenations reactions with OH (and O in nitrate hydrate), ending into CO_2 by reaction with OH (and O). The nitrate photolysis strongly increases the methanol destruction through the release of O (a factor of 10 at the studied concentration). This show how important can be the effect of the reactive oxygenated species. However, this cannot be generalized: glycine is an example where the OH radical plays no role. Finally, as shown above, NEXAFS in condensed films is a synchrotron based technique, which allows the study of the chemical reaction steps involved in the photochemistry of biologically relevant molecules. The role of the chemical environment (water and oxygen in particular) can be investigated.

Acknowledgments

The authors thank K. Shulte, J. N. Andersen, F. Sirotti and M. Silly for their assistance during the experiments.

References

[1] Fujii K, Shikazono N and Yokoya A 2009 *J. Phys. Chem. B* **113** 16007.
[2] Lacombe S and Le Sech C 2009 *Surf. Sci.* **603** 1953.
[3] Laffon C, Lacombe S, Bournel F and Parent Ph 2006 *J. Chem. Phys.* **125** 204714.
[4] Lacombe S, Bournel F, Laffon C and Parent Ph 2006 *Angw. Chem. Int. Ed.* **45** 4159.
[5] Parent Ph, Bournel F, Lasne J, Lacombe S, Strazzula G, Gardonio S, Lizzit S, Kappler J-P, Joly L, Laffon C and Carniato S 2009 *J. Chem. Phys.* **131** 154308.
[6] McLaren R, Clark S A C, Ishii I, Hitchcock A P 1987 *Phys. Rev. A* **36** 1683.
[7] Jamieson C S, Mebel A M, Kaiser R I 2006 *Astrophys. J. Suppl. Ser.* **163** 184.
[8] Watanabe N, Kouchi A 2002 *Astrophys. J.* **567** 651.
[9] Luiti G, Dondes S, Herteck P 1966 *J. Chem. Phys.* **44** 405.
[10] Gerakines P A, Schutte W A, Ehrenfreund P 1996 *Astron. Astrophys.* **312** 289.
[11] Porter R P, Noyes W A 1958 *J. Am. Chem. Soc.* **88** 2310.
[12] Bennett C J, Chen S-H, Sun B-J, Chang H H A, Kaiser R I 2007 *Astrophys. J.* **660** 1588.
[13] Lepage M, Michaud M and Sanche L 1997 *J. Chem. Phys.* **107** 3478.
[14] Krim L, Lasne J, Laffon C and Parent Ph 2009 *J. Phys. Chem. A* **113** 8979.
[15] Fisher M and Warneck P J 1996 *Phys. Chem.* **100** 18745.
[16] Bock J and Jacobi H-W 2010 *J. Phys. Chem. A* **114** 1790.

[17] Zubavichus Y, Fuch O, Weinhart L, Heske C, Umbach E, Denlinger JD, Grunze M 2004 *Radiat. Res.* **161** 346.

[18] Kaneko F, Tanaka M, Narita S, Kitada T, Matsui T, Nakagawa K, Agui A, Fujii K and Yokoya A 2005 *J. Electron Spectrosc. Relat. Phenom.* **144** 291.

[19] Wilks R G, MacNaughton J B, Kraatz H –B, Regier T, Blyth R I R and Moewes A 2009 *J. Phys. Chem. A* **133** 5360.

[20] Gordon M L, Cooper G, Morin C, Araki T, Turci C C, Kaznatcheev K and Hitchcock A P 2003 *Phys Chem A* **107** 6144.

[21] Parent Ph, Laffon C, Mangeney C, Bournel F and Tronc M 2002 *J. Chem. Phys.* **117** 10842.

[22] Galano A, Alvarez-Idaboy J R, Montero L A and Vivier-Bunge A 2001 *J. Comput. Chem.* **22** 1138.

Femtosecond fluorescence studies of DNA/RNA constituents

T Gustavsson, A Banyasz, R Improta[1] and D Markovitsi
Laboratoire Francis Perrin, CEA/DSM/IRAMIS/SPAM - CNRS URA 2453
CEA/Saclay, 91191 Gif-sur-Yvette, France

E-mail : thomas.gustavsson@cea.fr

Abstract. In this overview, femtosecond fluorescence studies of various DNA constituents are presented, ranging from the monomeric chromophores to different model helices. In order to interpret the experimental results in terms of fundamental processes on the molecular scale they are discussed in the light of recent theoretical calculations. The ultrafast fluorescence decay observed for the monomers is explained by the involvement of highly efficient conical intersections (CI) between the first singlet excited state and the ground state. For the model helices, the picture is more complex, but fluorescence anisotropy data reveal collective effects.

1. Introduction

The interaction between irradiation and nucleic acids (DNA/RNA) is currently a subject of intense research. Direct irradiation-triggered as well as indirect radical-induced chemical alterations of the double helix may lead to the formation of mutations and ultimately cancer. A central but badly elucidated role in the chain of events leading up to the biological consequences is played by the initially electronically excited states of nucleic acids. These transient states have the hard task to handle the excess energy in such a way to minimize damage.

Since a few years we focus our research on the directly UV-excited electronic states in nucleic acids in order to characterize their structure and dynamics with regards to their reactivity. To probe the initially populated states, we have developed specific time-resolved fluorescence spectroscopic techniques covering a very large temporal domain, from ≈100 femtoseconds to the ≈100 nanoseconds.

Time-resolved fluorescence is a powerful tool to study molecular excited state dynamics. In principle, when the fluorescence arises from a well defined excited state, its intensity reflects directly the excited state population. In most molecules, and for nucleic acids in particular, the excited state population decay is ruled by non-radiative deactivation processes, so by measuring the fluorescence decay one characterizes the non-radiative mechanisms. In addition, valuable information can be obtained from the fluorescence anisotropy decays, which not only reflect the molecular reorientational dynamics, but also informs on ultrafast changes of the electronic structure. Comparison of the experimental results with theoretical models is necessary for the understanding of the underlying non-radiative processes.

However, DNA and RNA are very complex organized molecular systems even though composed of only four basic monomeric chromophores; two purines (adenine, A, and guanine, G) and two pyrimidines (cytosine, C and thymine/uracil, T/U), see figure 1. A huge amount of experimental and computational studies has thus been devoted during the last few years to the monomeric chromophores; the nucleic acid bases, nucleosides and nucleotides (NABs) of DNA/RNA. An

[1] Permanent address : Istituto Biostrutture e Bioimmagini – CNR, Via Mezzocannone 16, 80134 Napoli, Italy.

Journal of Physics: Conference Series **261** (2011) 012009 doi:10.1088/1742-6596/261/1/012009

important, but less abundant, number of studies treat simple model helices. Instructive overviews can be found in refs. [1-6].

In this article we will briefly overview recent time-resolved fluorescence studies of NABs in room-temperature solution performed with femtosecond time-resolution, either by fluorescence upconversion (FU) or by optical Kerr gating (OKG). We examine in more detail the behavior of the monomeric building blocks, whose fluorescence lifetime is extremely short and we tackle that of model helices. As a matter of fact, the fluorescence of model helices contains important nanosecond components and femtosecond measurements need to be combined with data obtained by other techniques, for example time-correlated single photon counting (TCSPC) [7-10]. However, this longer-lived fluorescence is outside the scope of the present article.

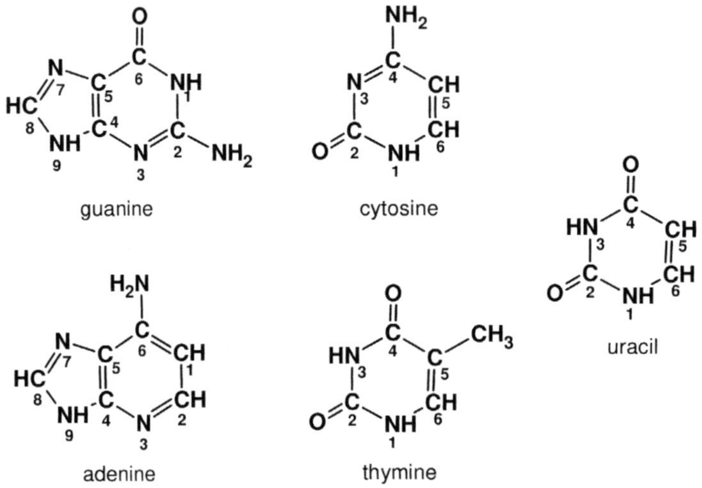

Figure 1. Structures of the most common naturally occurring nucleic acid bases.

2. Monomeric chromophores

Information regarding the initially excited Franck-Condon states can be deduced already from the absorption spectra which peak close to 260 nm for all the monomeric NABs. Analysis of the absorption spectra in combination with theoretical calculations indicate that for the pyrimidines there is only one strong $\pi\pi^*$ transition in the first UV band, while for the purines there are two close-lying $\pi\pi^*$ states, (L_b and L_a), which is clearly visible for guanine [1]. For all these heterocyclic molecules the possible presence of low-lying "dark" $n\pi^*$ states has played an important role in the discussions regarding the excited state dynamics.

Relying on the steady-state quantum yield measurements it was deduced already back in the 70's that the fluorescence lifetimes of the monomeric NABs in room-temperature water solution were on the order of a picosecond or less [11], implying extremely efficient internal conversion processes. Such short times eluded direct measurements for many years until femtosecond spectroscopic techniques became available.

During the last ten years, numerous femtosecond fluorescence studies have been performed on the monomeric chromophores in room-temperature water solution. These have been performed either by FU [12-20] or by OKG [21,22]. They all agree in assigning sub-picosecond lifetimes to the bright excited states. A comparison of the fluorescence decays for the four mononucleotides is given in figure 2.

We would like to stress the fact that the fluorescence decays of all NABs are strongly non-exponential, which can be seen in figure 2. This non-exponentiality indicates the presence of complex and/or multiple excited state relaxation processes. Moreover, there is a marked difference in the decay between the nucleoside and the nucleotide for the pyrimidines while it is negligible for the purines

[15]. This should be considered in view of the fact that the steady-state fluorescence spectra of nucleosides and nucleotides are identical. In the following we will rapidly look through the available experimental data without going into details, which will be further discussed below when reviewing the theoretical works.

The fluorescence decay of uracil is by far the fastest of the natural NABs [18]. In practice, its determination is limited by the experimental time-resolution, which is about 100 fs, meaning that it could be substantially faster. The fluorescence decays of the two other pyrimidines, thymine and cytosine, are substantially longer.

Concerning thymine [13,15,18,22], bi-exponential fitting gives an ultrafast component, ranging between 150 and 300 femtoseconds, followed by a slower component around one picosecond having much lower amplitude. In their broadband femtosecond Kerr-gate study of thymidine, Kwok *et al.* [22] assigned the origin of the bi-exponential decays to the existence of an intermediate state favoring the excited state deactivation. This so-called "doorway" state ("doorway" since it leads either to the ground state or the triplet state) could be the dark $^1n\pi^*$ state mixing with the close-lying $^1\pi\pi^*$ state.

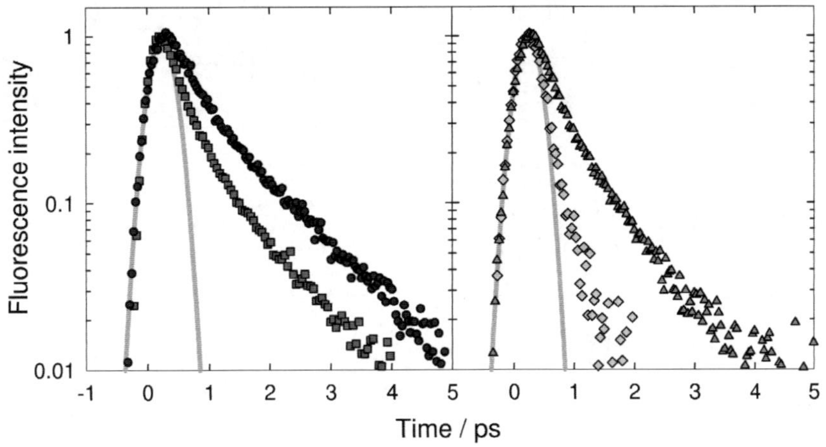

Figure 2. Total fluorescence decays at 330 nm for the four nucleotides, dCMP (red squares), TMP (blue circles) dAMP (green diamonds) and dGMP (orange triangles), after excitation at 267 nm. Also shown is the 330 fs (fwhm) instrument response function (gray).

Likewise, for cytosine, the fluorescence decays show an ultrafast component of 0.2-0.4 ps and a slower one [15,16,19]. This slower component is about one picosecond for the nucleoside and 1.4-1.7 ps for the nucleotide. Schwalb *et al.* [19] evoked different internal conversion pathways as the origin for the two decay times, the slower involving a dark $^1n\pi^*$ state.

The fluorescence decays of adenosine and adenosine monophosphate, recorded either by FU or OKG, are nearly as fast as that of uracil [12,14,15,21]. Actually, a bi-exponential fit gives a fast component of about 100 fs and longer one of a few hundred femtoseconds [15,21]. Kwok *et al.* associated the bi-exponentially with the L_a and L_b $^1\pi\pi^*$ excited states, respectively. The fast component should correspond to the L_a - L_b internal conversion. It is worth mentioning that the fluorescence decay of adenine is much slower than those of dA and dAMP, with a long component of about 8 ps and a faster one of a few hundred femtoseconds [14,17]. The fast component was assigned to the canonical 9H tautomer while the long one was assigned to the 7H tautomer. Adenine is the only NAB for which different tautomeric forms have been shown to be present in aqueous solution [23].

Regarding the guanine chromophore, the base itself (guanine) is not soluble enough in water so there are only data for the nucleoside and the nucleotide. It is characterized by a very broad steady-state fluorescence spectrum, extending beyond 700 nm [15]. Contrary to the other monomeric NABs,

Journal of Physics: Conference Series **261** (2011) 012009 doi:10.1088/1742-6596/261/1/012009

the fluorescence decays depend strongly on the wavelength, indicating important spectral shifts [20]. For dGMP, the fast component varies between 0.16 and 0.29 ps and the slow one between 1 and 4 ps going from 310 to 600 nm. This cannot be explained by solvation dynamics, but rather as due to diffusive wave packet dynamics on a flat excited state surface, in accordance with recent theoretical calculations [24]. (See also below.)

One may resume the results on the NABs as follows. The fluorescence decays are clearly non-exponential but can be approximately described by an ultrafast component (≈100 fs) and a slower one (ranging from about 0.4 ps for 2'-deoxyadenosine up to 1.7 ps in the case of cytidine monophosphate [19]). It is worthwhile to mention that there are no "long" components (> 3 ps) in the fluorescence decays of monomeric NABs in water solution. As described below, much longer fluorescence decays have been observed in chloroform solution. [25] A brief review of our results on NAB fluorescence lifetimes in water is given in Table 1.

Table 1. Fluorescence lifetimes (ps) of NABs measured at 330 nm in aqueous solution. They were obtained from fits with bi-exponential functions; $\alpha\exp(-t/\tau_1) + (1-\alpha)\exp(-t/\tau_2)$

Compound	τ_1	τ_2	α	$<\tau>$ [a]	ref.
Ura	0.10±0.01[b]				[18]
Thy	0.20±0.02	0.63±0.02	0.56±0.02	0.39 ± 0.02	[18]
dT	0.15±0.02	0.72±0.03	0.70±0.02	0.32 ± 0.01	[15]
TMP	0.21±0.03	1.07±0.06	0.67±0.02	0.50 ± 0.02	[15]
Cyt	0.20±0.02	1.30±0.07	0.85±0.02	0.37 ± 0.01	[16]
dC	0.18±0.02	0.92±0.06	0.83±0.02	0.30 ± 0.01	[15]
dCMP	0.27±0.02	1.38±0.11	0.84±0.02	0.45 ± 0.02	[15]
Ade	0.23±0.05	8.0±0.3	0.65 ± 0.05	2.95	[14]
dA	0.10±0.01[b]	0.42 ± 0.10	0.91 ± 0.01	0.13 ± 0.01	[15]
dAMP	0.10±0.01[b]	0.52 ± 0.10	0.94 ± 0.02	0.13 ± 0.01	[15]
dG	0.16±0.02	0.78±0.05	0.73±0.02	0.33 ± 0.01	[15]
dGMP	0.16±0.02 - 0.29±0.08[c]	0.94±0.09 - 4.0±1.0[c]			[20]

[a)] $<\tau> = \alpha\tau_1 + (1 - \alpha)\tau_2$,
[b)] limited by the time-resolution after deconvolution,
[c)] varying with the emission wavelength (310 - 600 nm).

The central question is of course what mechanism is responsible for the ultrafast internal conversion from the bright $^1\pi\pi^*$ state to the ground state for each of the monomeric NABs. Are there common features or are the processes involved unique for each molecule? In order to elucidate these questions, experimental studies of substituent and solvent effects in combination with high-level quantum chemistry calculations have proven to be essential.

2.1. Substitution effects

In addition to the studies of the naturally occurring RNA/DNA bases described above, various substituted bases have been characterized by femtosecond fluorescence spectroscopy. Among all the possible side group substitutions, methylation is particularly important since some types of methylation occur naturally.

Replacing the hydrogen in position 5 on uracil by a methyl group, as is the case of thymine, increases the fluorescence lifetime from less than 100 fs to about 400 fs [18]. This increase may seem minor, but substituent effects are of course very complex, not only the local chemical bond is changed but the electronic structure of the excited state may differ. Another and very particular case of methylation is 5-methyl-2'-deoxycytidine (5MdC), which is known to play a key role in biological functions [26]. Compared to natural cytosine, methylation causes a huge increase in fluorescence lifetime. The mean decay time for the 5MdC excited state is about one order of magnitude longer than for dC [27]. Interestingly, the fluorescence decays of 5MdC show strong wavelength dependence, which indicates complex excited state dynamics involving multiple states and/or species.

The effect of methylation is illustrated in figure 3 where the fluorescence decays at 330 nm of uracil and thymine as well as dC and 5MdC are shown. The decay of uracil is limited by the instrument response function while the two traces of thymine and dC are practically identical. The fluorescence decay of 5MdC is by far the longest.

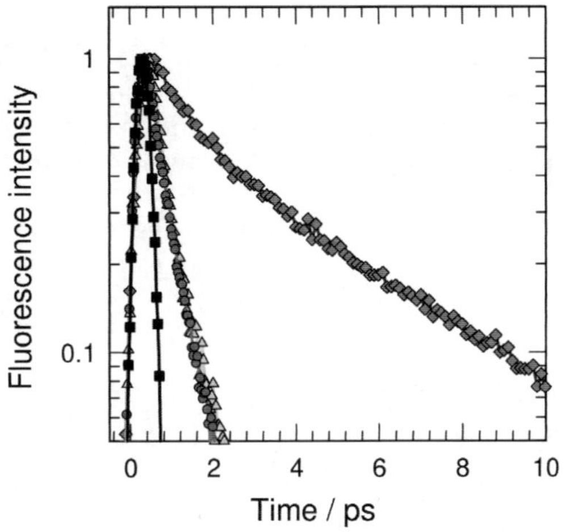

Figure 3. Methylation effect; uracil (black squares) and thymine (red circles), dC (green triangles) and 5MdC (pink diamonds) at 330 nm.

Performing systematic studies of substituent effects is therefore a very powerful tool to gather information about the electronic structure of the excited state that can be compared with predictions from quantum chemistry calculations. (see below)

We have performed such a systematic study of substituent effects in uracil [18,28-30], by using methyled, halogenated uracils and other modifications. Methylation at positions 1, 3 or 6 has no observable effect on the fluorescence lifetime; the fluorescence decays for these compounds are as fast as for uracil itself. Only methylation at position 5 (5-methyluracil = thymine) has significant effect, increasing the average lifetime by a factor of four. Moreover, it was found that the nature of the substituent in the 5-position influences strongly the fluorescence lifetime. For 5-fluorouracil, a mean lifetime of 1.3 ps is observed, more than ten times longer than that of uracil itself. Based on this study, it was shown that the excited state deactivation mechanism of uracil involves an important displacement of the 5-substituent. In parallel, a picosecond fluorescence study of a "blocked" uracil, 5,6-trimethyleneuracil, showed that such geometrical constraints has a dramatic effect, leading to a substantially longer fluorescence lifetime [31,32]. The authors proposed that the reaction path involves motion of both the 5- and the 6-substituent.

As in the case of uracils, the nature and position of substituents may strongly influence the electronic structure of the excited states and thus the fluorescence properties also for other NABs. In this context, the amino group has a particular role in the structure and dynamics of the NABs. It is

Journal of Physics: Conference Series **261** (2011) 012009 doi:10.1088/1742-6596/261/1/012009

naturally present in the cytosine, adenine and guanine chromophores, but not in uracil/thymine. Still all these chromophores are characterized by very short fluorescence lifetimes. What is more important is that the actual position of the amino group strongly influences the excited state properties as shown in the case of two aminouracils [33]. Both 5- and 6-aminouracil have much red-shifted absorption and fluorescence spectra, indicating that the amino-group influences the electronic structure. Interestingly, amino substitution on the 6-position does not affect the lifetime, which remains as short as for uracil, while amino substitution on the 5-position leads to a significant increase in lifetime. Moreover, the fluorescence decays of 5AU depend strongly on the emission wavelength, indicating a complex excited state relaxation mechanism[34].

As described above, the adenine chromophore (6-aminopurine) has a very short fluorescence lifetime while in the case of N,N-dimethyladenine the fluorescence is much longer-lived and extends into the visible region. A global analysis of fluorescence upconversion data produced 5 different time constants ranging between 100 fs and 62 ps [35,36]. These very complex dynamics were interpreted in terms of a 4-state model including $^1\pi\pi^*$ (L_a), $^1\pi\pi^*$ (L_b), $^1n\pi^*$ as well as an intramolecular charge-transfer state.

2-aminopurine (2AP), which is a structural analog to adenine, is highly fluorescent with a 11.8 ns lifetime [37]. 2AP was studied in water and in ethanol by FU, indicating important dynamical spectral shifts which were assigned to solvation dynamics [38]. More interestingly, 2-aminopurine can form a Watson-Crick base pair with thymine and thus be incorporated naturally in a DNA double helix and is used as a fluorescent probe to follow its conformational dynamics on longer times [39]. However, ultrafast charge transfer in 2AP modified DNA sequences has been observed by FU [40]. The 2AP fluorescence and fluorescence anisotropy have also been used to follow rapid RNA conformational dynamics [41].

2.2. *Solvent effects*

We now turn to the issue of solvent effects. With this we understand local environmental effects on the fluorescence properties of the monomeric NABs. This issue is not limited to the understanding of the photophysics of individual NABs but is related to a more fundamental question; to what extent the electronic structure of UV excited DNA can be considered "monomeric". For a given monomeric chromophore, the surrounding bases in the DNA helix may constitute a "solvent". Of course, this local environment is very different from water, the solvent in which the large majority of spectroscopic investigations of NABs have been performed. For these reasons, it is of great interest to characterize the NAB fluorescence in different solvents of different polarity, hydrogen bonding capacity and viscosity.

Solvent effects may be "static", in the sense that the solvent influences the energetic ordering of the molecular excited states, but they may also be dynamic, i.e. how the dipolar solute-solvent interaction evolves with solvent rearrangement. Based on the fact that the NABs excited state lifetimes are probably faster than the solvent response, the dynamic effect of the solvent on the singlet excited state relaxation processes was up to recently thought to be "quite modest" [1]. This may be true for "slower" solvents such as alcohols, while for water the solvation dynamics are probably too fast for the time-resolution available with presently used experimental setups. Solvation dynamics in water is dominated by an ultrafast "inertial" component of only a few tens of femtoseconds, although it also contains a slower picosecond component of low amplitude [42]. Indeed, spectral shifts due to solvation dynamics have not been reported for any of the natural NABs in water.

For some modified bases, for example 5MdC [27] and 5AU [33], which have much longer excited state lifetimes, important dynamical spectral shifts have been observed in water, but it is today not clear to what extent these may be linked to the solvent response. It is rather believed that they are due to intramolecular processes. A clear example is the case of 2-aminopurine [38], for which dynamical spectral shifts (0.2 and 0.9 ps in water) were observed and assigned to solvation dynamics.

However, the solvent does not only influence the excited state dynamics but may also directly affect their relative energy ordering (static effect), and thus radically change the photophysical properties of the chromophore.

In our group, we have studied how the fluorescence properties of chosen uracils depend on the solvent [28-30,43]. An example is given in figure 4 where the fluorescence decays of thymine and 5-fluorouracil in acetonitrile, methanol and water are compared.

As can be seen in figure 4, there are significant differences in decay times for a given chromophore in different solvents. The decays observed in aprotic acetonitrile are the fastest while those in protic water are the slowest. This phenomenon cannot be explained in simple terms of macroscopic solvent properties such as polarity or viscosity, but is instead related to how the solute-solvent interaction (dipolar, H-bonding) fine-tunes the energetic ordering of the chromophore's excited states. Briefly, the relative positioning of a "dark" $n\pi^*$ state with regards to the bright $\pi\pi^*$ state may vary with the solvent. If it is situated below the $\pi\pi^*$ state it will accelerate the apparent deactivation rate for which the internal conversion rate is already very fast.

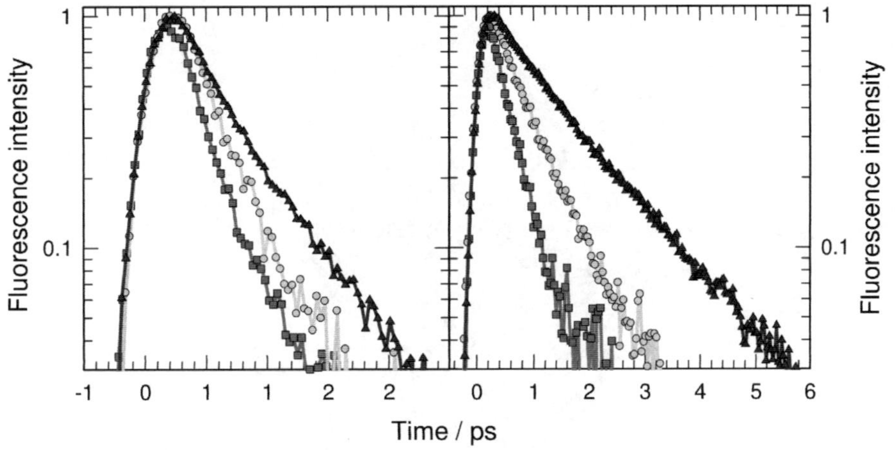

Figure 4. Fluorescence decays at 330 nm of thymine (left) and 5-fluorouracil (right) in acetonitrile (red squares), methanol (green circles) and water (blue triangles) after excitation at 267 nm.

In a fluorescence upconversion study of modified cytidine and guanosine in chloroform solution after excitation at 283 nm, Schwalb and Temps measured much more complex decays than what was observed in water [25]. For both compounds, ultrafast components comparable to those in water, were observed but also much longer ones; for cytidine 21 ps and for guanosine 7 ps as well as low amplitude component of hundreds of picoseconds. The long-lived components were assigned to $n\pi^*$ states, since these are expected to be more stable than the directly excited $\pi\pi^*$ state in chloroform. [25] As these molecules were subtituted with a tert-butyldimethylsilyl group instead of the deoxyribose group to render them soluble in chloroform, it is not straightforward to compare these results with those obtained in aqueous solution using natural cytidine and guanosine.

In their studies of N,N-dimethyladenine, Schwalb and Temps compared the excited state charge transfer dynamics in water to those in dioxane [35,36]. The much longer lifetime of the ICT state in dioxane (1.4 ns) than that in water (62 ps) was taken as an indication that the deactivation mechanism is strongly affected by the solvent.

The very low fluorescence quantum yields and the ultrafast fluorescence decays of the NABs imply the presence of highly efficient nonradiative deactivation processes taking place in the first excited singlet state. A deeper understanding of these processes can only be obtained through high-level theoretical calculations. Consequently, many theoretical studies based on quantum chemistry

Journal of Physics: Conference Series **261** (2011) 012009 doi:10.1088/1742-6596/261/1/012009

calculations have been dedicated to uracil/thymine [18,31,44-55], cytosine [32,47,56-67], adenine [46,51,68-78] and guanine [24,79-86].

2.3. The non-radiative deactivation mechanism

The picture emerging is that the ultrafast decay of the excited state is due to highly efficient conical intersections (CI) between the first singlet excited state and the ground state. These CIs are related to conformational changes taking place in the "bright" $^1\pi\pi^*$ state. Obviously, the geometrical changes involved in the CIs differ from one molecule to the other. As indicated above, a complication is the fact that the deactivation processes may involve near-lying "dark" state(s), for example of $^1n\pi^*$ character, which are basically inaccessible by fluorescence spectroscopy. Moreover, most theoretical work has been performed in vacuum, while this article focuses on solution phase fluorescence. Since the relative energy ordering of the various excited states is very sensitive to the environment, a direct comparison between theory and experiment is not always pertinent.

From a theoretical point of view, an instructive comparison of the nonradiative processes involved in the excited state deactivation for the NABs has recently been given by Serrano-Andres and Merchan [87].

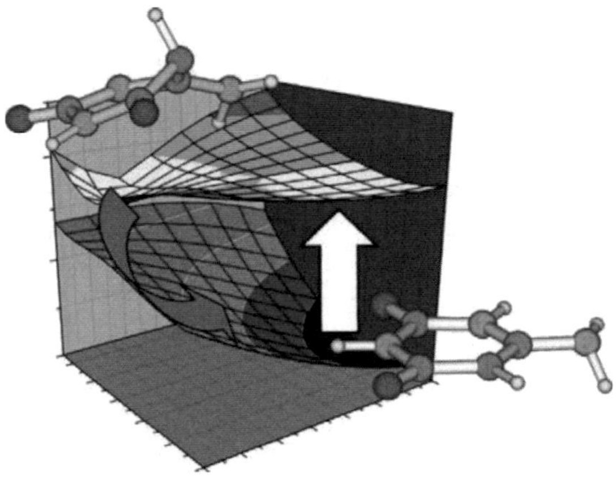

Figure 5. Proposed deactivation mechanism for uracil derivatives. The figure depicts actually 6-aminouracil, for which more details can be found in ref. [34]

Generally speaking, one may say that for the pyrimidines, the internal conversion through the CI involves a pyramidalization of C_5 and a torsion of the C_5C_6 bond associated with an out-of-plane motion of the 5-substituent. This is valid both for uracil/thymine [18,48] and cytosine [65]. In figure 5 this reaction is schematically shown for 6-aminouracil. However, alternative quantum chemistry calculations indicate that not only the 5-substituent but also the 6-substituent are moving out of the molecular plane forming a biradical state [31,32].

For the purines, on the other hand, the internal conversion from the first excited $^1\pi\pi^*$ state of the ground state involves a twisting around the C_2N_3 bond provoking an out-of-plane bending of the 2-substituent. For guanine the CI to the ground state can be described as an out-of-plane distortion of the C_2 center [83]. The observed bi-exponential excited state dynamics has been attributed to the wavepacket motion along a barrierless pathway on a "flat" $^1\pi\pi^*$ state surface towards the CI [24]. Also for adenine calculations indicate that the CI is reached via a relatively flat energy region before the hydrogen atom in the 2-position moves out of the molecular plane [78].

Such a general description is bound to be oversimplified, and a very active debate exists regarding the details of the decay mechanism for each individual base. It is important to underline that eventhough the main features seem to be well understood, many aspects still need further theoretical considerations, such as the actual shape of the excited state potential energy surface along the reaction coordinate towards the CI. Needless to say, a correct description can only be obtained by correctly incorporating solvent effects. Finally, to describe the excited state evolution quantitatively and thus enable comparisons with measured lifetimes, quantum dynamics calculations will be required. Such computations are beginning to appear only now [49,55,66,88,89].

3. DNA helices

The time-evolution of the fluorescence of DNA model helices [7-10,90-94] as well as that of natural DNA [95] is different from those of their monomeric constituents. This concerns in particular the fluorescence lifetimes but in many cases also the fluorescence spectra. At this point, let us rapidly side-step a bit to note that already from steady-state absorption spectra there is strong evidence that the first UV band corresponds to collective excitations, i.e. excitonic states that are populated directly by photon absorption [6,96]. The processes involved in the time evolution of these delocalized states and the nature of the emitting state(s) are the subjects of a very intense debate today [5,97]. It is not our purpose here to make a complete "synthesis" of the time-resolved fluorescence properties of the various model helices studied to this date. The main findings are that the fluorescence decays, and thus non-radiative relaxation mechanisms active in the excited state, are strongly sequence dependent.

Already from the first femtosecond fluorescence studies it was obvious that the fluorescence lifetimes, and thus the excited state lifetimes, were very different for the organized systems with regards to the monomeric NABs. For the adenine and thymine containing single and double strands $(dA)_{20}$, $(dT)_{20}$ and $(dA)_{20}\cdot(dT)_{20}$, fluorescence decays at 330 nm showed picosecond lifetimes [7,92], which is much longer than those of the monomers. Interestingly, for the alternating oligomer $(dAdT)_{10}\cdot(dAdT)_{10}$ a much shorter lifetime was observed at this wavelength compared to that of the homopolymeric duplex [93]. Such sequence effects are also observed for polymeric duplexes [97], as illustrated in figure 6.

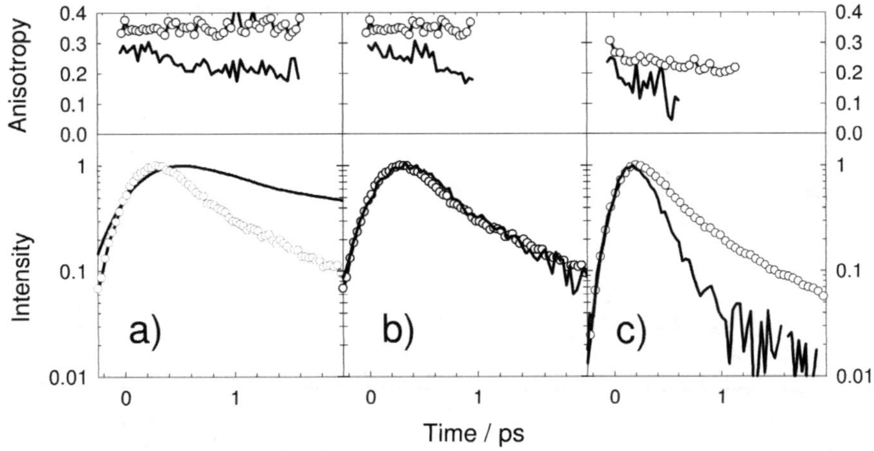

Figure 6. Fluorescence anisotropies (upper) and fluorescence decays (lower) obtained at 330 nm for a) poly(dA)•poly(dT), b) poly(dAdT)•poly(dAdT) and c) poly(dGdC)•poly(dGdC) in phosphate buffer. Also shown are the corresponding traces for an equi-molar mixture of monomers (circles).

The fluorescence decays recorded at 330 nm of poly(dGdC)·poly(dGdC), poly(dAdT)·poly(dAdT) and poly(dA)·poly(dT) are successively shorter, equal and longer than those of equimolar mixtures of the constitutive nucleotides. The extraordinary fast decay observed for poly(dGdC)·poly(dGdC) [98] is in line with experiments on isolated Watson-Crick guanine-cytosine pairs dissolved in chloroform [25]. More generally speaking, the presence of guanine-cytosine pairs in homopolymeric purine-pyrimidine oligomeric duplexes leads to a decrease of the FU lifetimes [99]. Theoretical calculations emphasized the role of interbase (interstrand) proton transfer in the internal conversion mechanism involving the formation of an excited charge transfer state [100-102].

However, it is worth noticing that fluorescence lifetimes measured by TCSPC are longer than those obtained by transient absorption spectroscopy [1,2,8,9,91,94,95].

Sequence effects are thus clearly established, but of course there is no straightforward explanation of the observed fluorescence lifetimes in terms of this single variable. The excited state non-radiative processes active in helices are very complex and depend on many other factors as well. However, additional information on the excited states has been obtained from fluorescence anisotropy measurements. For all model helices discussed here a sub-picosecond decay of the anisotropy is observed, as shown in figure 6. Such an ultrafast decay cannot be due to a physical rotation of the helix and is instead interpreted in terms of energy transfer within the excitonic band which is initially delocalized over several base pairs [5,6].

Single strands containing guanine runs have the propensity to self-associate into four-stranded structure called quadruplexes. To this date, only one system characterized by a well-defined quadruplex structure, formed by four thymine-capped single strands d(TGGGGT), has been studied with femtosecond resolution [103]. The fluorescence lifetime of these quadruplexes were found to be much longer than that of the corresponding monomeric chromophores. Although sequence effects on the lifetimes of the bright $\pi\pi^*$ of model DNA helices have been clearly established many questions remain open. These have been outlined in ref. [10].

4. Conclusion

To conclude, we have shown how femtosecond spectroscopic techniques have contributed to elucidate the mechanisms involved in the very efficient excited state deactivation of nucleic acids, both monomers and model helices. In order to truly understand the underlying mechanisms, experimental results need to be combined with theoretical calculations, which are today only feasible on the monomeric level. Finding a correct theoretical description of the excited states of model helices constitutes a major challenge in the years to come.

References

[1] Crespo-Hernández C E, Cohen B, Hare P M and Kohler B 2004 *Chem. Rev.* **104** 1977.
[2] Middleton C T, de La Harpe K, Su C, Law Y K, Crespo-Hernandez C E and Kohler B 2009 *Ann. Rev. Phys. Chem.* **60** 217.
[3] Kohler B 2010 *J. Phys. Chem. Letts.* **1** 2047.
[4] Gustavsson T, Improta R and Markovitsi D 2010 *J. Phys. Chem. Letts.* **1** 2025.
[5] Markovitsi D 2009 *Pure Appl. Chem.* **81** 1635.
[6] Markovitsi D, Gustavsson T and Banyasz A 2010 *Mutat. Res. Rev.* **704** 21.
[7] Markovitsi D, Sharonov A, Onidas D and Gustavsson T 2003 *ChemPhysChem* **3** 303.
[8] Markovitsi D, Gustavsson T and Talbot F 2007 *Photochem. Photobiol. Sci.* **6** 717..
[9] Vayá I, Changenet-Barret P, Gustavsson T, Zikich D, Kotlyar A B and Markovitsi D 2010 *Photochem. Photobiol. Sci.* **9** 1193.
[10] Markovitsi D, Gustavsson T and Vayá I 2010 *J. Phys. Chem. Letts.* in press.
[11] Daniels M and Hauswirth W 1971 *Science* **171** 675.
[12] Peon J and Zewail A H 2001 *Chem. Phys. Lett.* **348** 255.
[13] Gustavsson T, Sharonov A and Markovitsi D 2002 *Chem. Phys. Lett.* **351** 195.
[14] Gustavsson T, Sharonov A, Onidas D and Markovitsi D 2002 *Chem. Phys. Lett.* **356** 49.
[15] Onidas D, Markovitsi D, Marguet S, Sharonov A and Gustavsson T 2002 *J. Phys. Chem. B* **106** 11367.
[16] Sharonov A, Gustavsson T, Carré V, Renault E and Markovitsi D 2003 *Chem. Phys. Lett.* **380** 173.

[17] Pancur T, Schwalb N K, Renth F and Temps F 2005 *Chem. Phys.* **313** 199.
[18] Gustavsson T, Banyasz A, Lazzarotto E, Markovitsi D, Scalmani G, Frisch M J, Barone V and Improta R 2006 *J. Am. Chem. Soc.* **128** 607.
[19] Schwalb N K and Temps F 2009 *J. Photochem. Photobiol. A-Chem.* **208** 164.
[20] Miannay F-A, Gustavsson T, Banyasz A and Markovitsi D 2010 *J. Phys. Chem. A* **114** 3256.
[21] Kwok W-M, Ma C and Phillips D L 2006 *J. Am. Chem. Soc.* **128** 11894.
[22] Kwok W-M, Ma C and Phillips D L 2008 *J. Am. Chem. Soc.* **130** 5131.
[23] Wilson R W and Callis P R 1980 *Photochem. Photobiol.* **31** 323.
[24] Karunakaran V, Kleinermanns K, Improta R and Kovalenko S A 2009 *J. Am. Chem. Soc.* **131** 5839.
[25] Schwalb N K and Temps F 2007 *J. Am. Chem. Soc.* **129** 9272.
[26] Dean W, Santos F, Stojkovic M, Zakhartchenko V, Walter J, Wolf E and Reik W 2001 *Proc. Natl. Acad. Sci. U.S.A.* **98** 13734.
[27] Sharonov A, Gustavsson T, Marguet S and Markovitsi D 2003 *Photochem. Photobiol. Sci.* **2** 1.
[28] Gustavsson T, Sarkar N, Lazzarotto E, Markovitsi D, Barone V and Improta R 2006 *J. Phys. Chem. B* **110** 12843.
[29] Gustavsson T, Sarkar N, Lazzarotto E, Markovitsi D and Improta R 2006 *Chem. Phys. Lett.* **429** 551.
[30] Gustavsson T, Sarkar N, Bányász Á, Markovitsi D and Improta R 2007 *Photochem. Photobiol.* **83** 595.
[31] Zgierski M Z, Patchkovskii S, Fujiwara T and Lim E C 2005 *J. Phys. Chem. A* **109** 9384.
[32] Zgierski M Z, Fujiwara T, Kofron W G and Lim E C 2007 *Phys. Chem. Chem. Phys.* **9** 3206.
[33] Bányász Á, Gustavsson T, Keszei E, Improta R and Markovitsi D 2008 *Photochem. Photobiol. Sci.* **7** 765.
[34] Bányász Á, Karpati S, Mercier Y, Reguero M, Gustavsson T, Markovitsi D and Improta R 2010 *J. Phys. Chem. B* **114** 12708.
[35] Schwalb N K and Temps F 2006 *Phys. Chem. Chem. Phys.* **8** 5229.
[36] Schwalb N K and Temps F 2009 *J. Phys. Chem. A* **113** 13113.
[37] Holmén A, Nordén B and Albinsson B 1997 *J. Am. Chem. Soc.* **119** 3114.
[38] Pal S K, Peon J and Zewail A H 2002 *Chem. Phys. Lett.* **363** 57
[39] Rachofsky E L, Seibert E, Stivers J T, Osman R and Ross J B A 2001 *Biochem.* **40** 957
[40] Manoj P, Min C K, Aravindakumar C T and Joo T 2008 *Chem. Phys.* **352** 333.
[41] Zhao L and Xia T 2007 *J. Am. Chem. Soc.* **129** 4118.
[42] Jimenez R, Fleming G R, Kumar P V and Maroncelli M 1994 *Nature* **369** 471.
[43] Gustavsson T, Banyasz A, Sarkar N, Markovitsi D and Improta R 2008 *Chem. Phys.* **350** 186.
[44] Improta R and Barone V 2004 *J. Am. Chem. Soc.* **126** 14320.
[45] Matsika S 2004 *J. Phys. Chem. A* **108** 7584.
[46] Matsika S 2005 *J. Phys. Chem. A* **109** 7538.
[47] Merchán M, Gonzalez-Luque R, Climent T, Serrano-Andres L, Rodriguez E, Reguero M and Pelaez D 2006 *J. Phys. Chem. B* **110** 26471.
[48] Santoro F, Barone V, Gustavsson T and Improta R 2006 *J. Am. Chem. Soc.* **128** 16312.
[49] Hudock H R, Levine B G, Thompson A L, Satzger H, Townsend D, Gador N, Ullrich S, Stolow A and Martinez T J 2007 *J. Phys. Chem. A* **111** 8500.
[50] Mercier Y, Santoro F, Reguero M and Improta R 2008 *J. Phys. Chem. B* **112** 10769.
[51] Improta R and Barone V 2008 *Theor.Chem. Acc.* **120** 491.
[52] Epifanovsky E, Kowalski K, Fan P D, Valiev M, Matsika S and Krylov A I 2008 *J. Phys. Chem. A* **112** 9983.
[53] Zechmann G and Barbatti M 2008 *J. Phys. Chem. A* **112** 8273.
[54] Improta R and Barone V 2009 *Theochem-J. Mol. Struct.* **914** 87.
[55] Improta R, Barone V, Lami A and Santoro F 2009 *J. Phys. Chem. B* **113** 14491.

[56] Ismail N, Blancafort L, Olivucci M, Kohler B and Robb M A 2002 *J. Am. Chem. Soc.* **124** 6818.
[57] Merchán M and Serrano-Andrés L 2003 *J. Am. Chem. Soc.* **125** 8108.
[58] Blancafort L and Robb M A 2004 *J. Phys. Chem. A* **108** 10609.
[59] Blancafort L, Cohen B, Hare P M, Kohler B and Robb M A 2005 *J. Phys. Chem. A* **109** 4431.
[60] Zgierski M Z, Patchkovskii S and Lim E C 2005 *J. Chem. Phys.* **123** 081101.
[61] Zgierski M Z and Alavi S 2006 *Chem. Phys. Lett.* **426** 398.
[62] Blancafort L 2007 *Photochem. Photobiol.* **83** 603.
[63] Kistler K A and Matsika S 2007 *J. Phys. Chem. A* **111** 2650.
[64] Blancafort L and Migani A 2007 *J. Photochem. Photobiol. A-Chem.* **190** 283.
[65] Kistler K A and Matsika S 2008 *J. Chem. Phys.* **128** 215102.
[66] Hudock H R and Martinez T J 2008 *ChemPhysChem* **9** 2486.
[67] Kistler K A and Matsika S 2010 *Phys. Chem. Chem. Phys.* **12** 5024.
[68] Marian C M 2005 *J. Chem. Phys.* **122** 104314.
[69] Nielsen S B and Sølling T I 2005 *ChemPhysChem* **6** 1276.
[70] Perun S, Sobolewski A L and Domcke W 2005 *Chem. Phys.* **313** 107.
[71] Perun S, Sobolewski A L and Domcke W 2005 *J. Am. Chem. Soc.* **127** 6257.
[72] Blancafort L 2006 *J. Am. Chem. Soc.* **128** 210.
[73] Serrano-Andres L, Merchán M and Borin A C 2006 *Chem. Eur. J.* **12** 6559.
[74] Barbatti M and Lischka H 2007 *J. Phys. Chem. A* **111** 2852.
[75] Chung W C, Lan Z, Ohtsuki Y, Shimakura N, Domcke W and Fujimura Y 2007 *Phys. Chem. Chem. Phys.* **9** 2075.
[76] Yamazaki S and Kato S 2007 *J. Am. Chem. Soc.* **129** 2901.
[77] Marian C M, Kleinschmidt M and Tatchen J 2008 *Chem. Phys.* **347** 346.
[78] Conti I, Garavelli M and Orlandi G 2009 *J. Am. Chem. Soc.* **131** 16108.
[79] Langer H and Doltsinis N L 2004 *Phys. Chem. Chem. Phys.* **6** 2742.
[80] Langer H, Doltsinis N L and Marx D 2005 *ChemPhysChem* **6** 1734.
[81] Chen H and Li S 2006 *J. Chem. Phys.* **124** 154315.
[82] Zgierski M Z, Patchkovskii S, Fujiwara T and Lim E C 2007 *Chem. Phys. Lett.* **440** 145.
[83] Marian C M 2007 *J. Phys. Chem. A* **111** 1545.
[84] Serrano-Andres L, Merchán M and Borin A C 2008 *J. Am. Chem. Soc.* **130** 2473.
[85] Yamazaki S, Domcke W and Sobolewski A L 2008 *J. Phys. Chem. A* **112** 11965.
[86] Lan Z G, Fabiano E and Thiel W 2009 *ChemPhysChem* **10** 1225.
[87] Serrano-Andrés L and Merchán M 2009 *J. Photochem. Photobiol. C: Photohem. Rev.* **10** 21.
[88] Santoro F, Improta R and Barone V 2009 *Theoretical Chemistry Accounts* **123** 273.
[89] Improta R, Santoro F, Barone V and Lami A 2009 *J. Phys. Chem. A* **113** 15346.
[90] Markovitsi D, Onidas D, Gustavsson T, Talbot F and Lazzarotto E 2005 *J. Am. Chem. Soc.* **127** 17130.
[91] Markovitsi D, Talbot F, Gustavsson T, Onidas D, Lazzarotto E and Marguet S 2006 *Nature* **441** E7.
[92] Onidas D, Gustavsson T, Lazzarotto E and Markovitsi D 2007 *J. Phys. Chem. B* **111** 9644.
[93] Onidas D, Gustavsson T, Lazzarotto E and Markovitsi D 2007 *Phys. Chem. Chem. Phys.* **9** 5143.
[94] Vayá I, Miannay F-A, Gustavsson T and Markovitsi D 2010 *ChemPhysChem* **11** 987.
[95] Vayá I, Gustavsson T, Miannay F-A, Douki T and Markovitsi D 2010 *J. Am. Chem. Soc.* **132** 11834.
[96] Burin A L, Armbruster M E, Hariharan M and Lewis F D 2009 *Proc. Natl. Acad. Sci. U.S.A.* **106** 989.
[97] Markovitsi D and Gustavsson T, in I. Burghardt, V. May, D.A. Micha, E.R. Bittner (Eds.), *Energy Transfer Dynamics in Biomaterial Systems*. Springer Verlag, Berlin Heidelberg, 2009 p. 127.

[98] Miannay FA, Banyasz A, Gustavsson T and Markovitsi D 2007 *J. Am. Chem. Soc.* **129** 14574.
[99] Schwalb N K and Temps F 2008 *Science* **322** 243.
[100] Sobolewski A L and Domcke W 2004 *Phys. Chem. Chem. Phys.* **6** 2763.
[101] Sobolewski A L, Domcke W and Hättig C 2005 *Proc. Natl. Acad. Sci. U.S.A.* **102** 17903.
[102] Groenhof G, Schafer L V, Boggio-Pasqua M, Goette M, Grubmuller H and Robb M A 2007 *J. Am. Chem. Soc.* **129** 6812.
[103] Miannay F A, Banyasz A, Gustavsson T and Markovitsi D 2009 *J. Phys. Chem. C* **113** 11760.

Journal of Physics: Conference Series **261** (2011) 012010 doi:10.1088/1742-6596/261/1/012010

Radiation damage to DNA-protein complexes

M Spotheim-Maurizot[1] and M Davídková[2]
[1] Centre de Biophysiqe Moléculaire, CNRS, rue C. Sadron, 45071 Orléans, France -
[2] Dept. of Radiation Dosimetry, Nuclear Physics Institute ASCR, Na Truhlářce 39/64, 18086 Prague 8, Czech Republic.

E-mail: spotheim@cnrs-orleans.fr

Abstract. We review here the advances in understanding the effects of ionizing radiations on DNA, proteins and their complexes, resulting from the collaboration of the authors' teams. It concerns the preponderant indirect effect of low LET ionizing radiations, thus the attack of the macromolecules in aqueous solution by the most aggressive product of water radiolysis, the hydroxyl radical. A model of simulation of the reaction of these radicals with the macromolecules (called RADACK) was developed and was used for calculating the probabilities of damage of each constituent of DNA or proteins (nucleotide or amino-acid). The calculations allowed to draw conclusions from electrophoresis, mutagenesis, spectroscopic (fluorescence, circular dichroïsm) and mass spectrometry experiments. Thus we have shown that the extent and location of the lesions are strongly dependent on the 3D structure of the macromolecules, which in turns is modulated by their sequence and by the binding of some ligands. Molecular dynamics simulation completed our studies in showing the consequences of each lesion on the stability and structure of the proteins and their complexes with DNA.

1. Introduction

Ionizing radiations are damaging DNA, proteins and their complexes *in vivo* (in cells) and *in vitro* (in aqueous solution) via their direct and indirect effects. The direct effects occur via the ionization and excitation of the macromolecules, whereas the indirect ones occur via the products of ionization and excitation of the water molecules surrounding the macromolecules.

The proportion of the two types of effects depends on the LET of the radiation, the proportion of direct effects being higher for high LET radiation (alpha particles, carbon ions, etc). In the case of low LET radiations such as gamma rays the main damaging species induced by the irradiation of macromolecules in solution are the sparsely distributed hydroxyl radicals ($OH^{\cdot}$) produced by water radiolysis. The proximity of the radicals produced in the high density tracks of high LET particles favors their recombination and thus reduces the proportion of indirect effects.

1.1 DNA radiolysis

The hydroxyl radicals are damaging DNA by abstracting H atoms from the sugar moiety (deoxyribose) and bases or add to the double bonds of certain bases. The final consequence of these reactions is the formation of strand breaks (called frank strand breaks, FSB), abasic sites or modified sugar and bases (figure1). Part of the nucleotides with modified bases is alkali labile and therefore such lesions can be revealed as strand breaks in alkaline pH (called alkali revealed breaks, ARB).

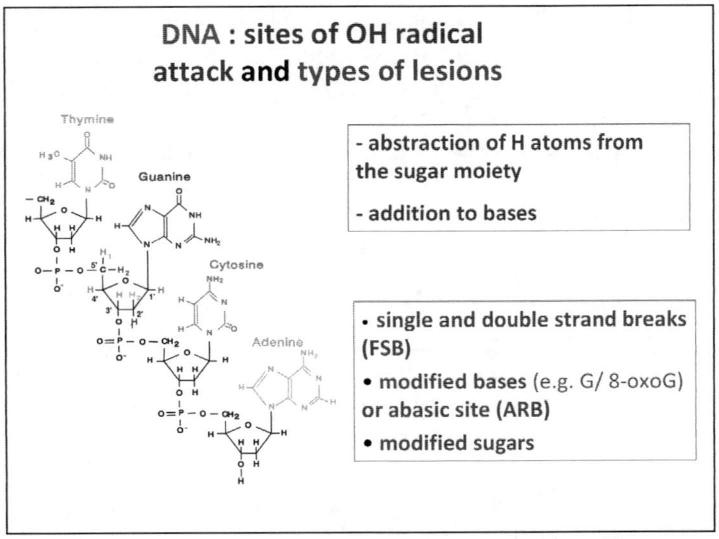

Figure 1. Attack of DNA by the OH· radicals produced by the radiolysis of water and its consequences.

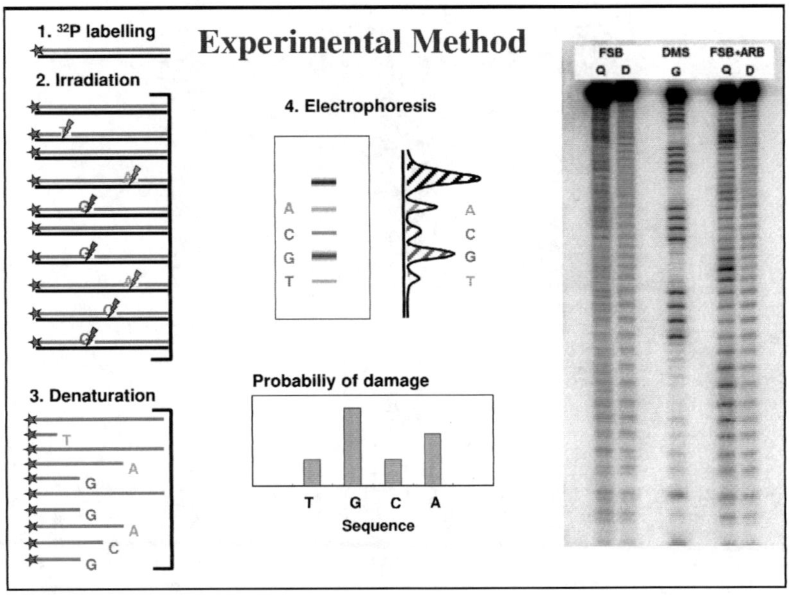

Figure 2. Sequencing gel electrophoresis applied to an irradiated double stranded (D) and quatruplex DNA (Q). 1. DNA is labeled with ^{32}P at the 5'- end of one strand. 2. DNA is irradiated and breaks occur on both strands. 3. DNA is denatured and radioactive fragments of different length are obtained. 4. The radioactive fragments are separated in a sequencing gel. The radioactivity of each band is quantified (area under the peak) using a PhosphoImager. It is proportional to the amount of radioactive fragments ending by a given (unlabelled) nucleotide. It is proportional to the probability of breakage at that nucleotide. On the example of gel, samples at neutral pH reveal FSB and samples incubated with piperidine (alkali) reveal FSB+ARB. The middle lane shows the bands corresponding to fragments ending by an unlabelled guanosine.

Their number and distribution along a DNA molecule can be determined by applying electrophoresis techniques. In the case of a circular supercoiled plasmid DNA, by taking into account that a single strand break converts the plasmid into a relaxed circular one, and that a double strand break leads to a linear one, the separation of the three forms by agarose gel electrophoresis allows to determine the number of single and double strand breaks [1]. In the case of a linear double stranded DNA, using sequencing gel electrophoresis of radioactively labeled irradiated fragments (figure 2) allowed to observe that the frequencies of FSB and ARB are not homogeneously distributed along the molecule, but are modulated by DNA sequence of nucleotides [2].

Using molecular modeling (figure 3) we were able to explain the observed radio-resistance of certain sequences of B-DNA (such as the 5'-AATT-3' sequence) by the reduced accessibility (defined as the surface of the van der Waals sphere representing the H atom on which a sphere of the size of an OH· radical can roll over) of the H4' and H5'2 atoms of sugars, atoms located in the narrow minor groove of such sequences3. The accessibility of the H5'1 atom situated at the edge of the minor groove is independent of the minor groove width.

It should be recalled that among all the H atoms of the sugar moiety, H4' and H5'2 are the most accessible to OH· radicals and their abstraction is the main cause of breakage of the phosphodiester backbone of DNA (by a beta-elimination reaction) [3].

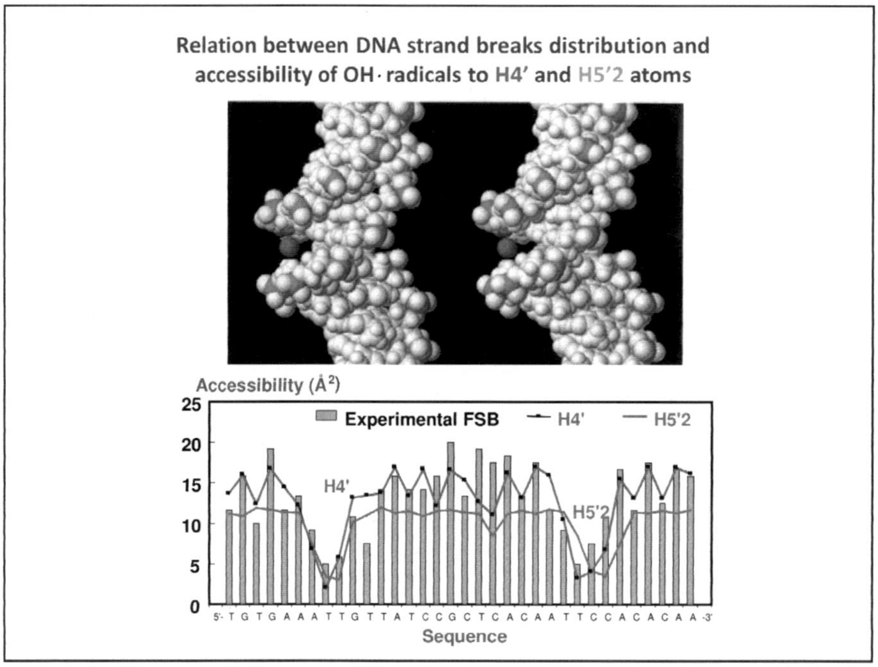

Figure 3. Accessibility of H4' and H5' atoms to OH· radicals in a B-DNA fragment. In the top picture: OH· radical in red, H4' in magenta, H5'1 in yellow and H5'2 in green.

Later on, an original model of simulation of radiolytic attack (RADACK) based on a Monte Carlo technique allowed to calculate the probability of breakage reactions at each nucleotide by taking into account not only the accessibility of attack sites, but also the chemical reactivity of the different bases and the efficiency with which the abstraction reaction leads to strand breakage [4]. The principle of the simulation model is presented in figure 4. The three dimensional structure of DNA was obtained either from structural databases such as PDB or was built by molecular modeling using the SYBYL software.

The same approach succeeded to account for the observed influence of more drastic variations of DNA structure (Z-DNA, quadruplexes) on the distribution of FSB and ARB [5-8]. Interestingly the efficiency with which the reaction of OH· radicals with sugar H atoms leads to strand breakage seems to be different in the left handed Z-form of DNA than in the canonical right handed B-DNA, probably because of differences in sugar conformation (for nucleotides containing guanines) and other structural parameters.

The consideration of local structural variations within studied biological macromolecules in the mentioned modeling approach is for the moment exclusive. Similar stochastic models usually apply track structures of incident radiation to calculate the spectrum of damages in large DNA fragments, chromosome domains or the whole cell nucleus, where biomolecules are represented by standard "average" molecular conformation (e.g. [9-10]).

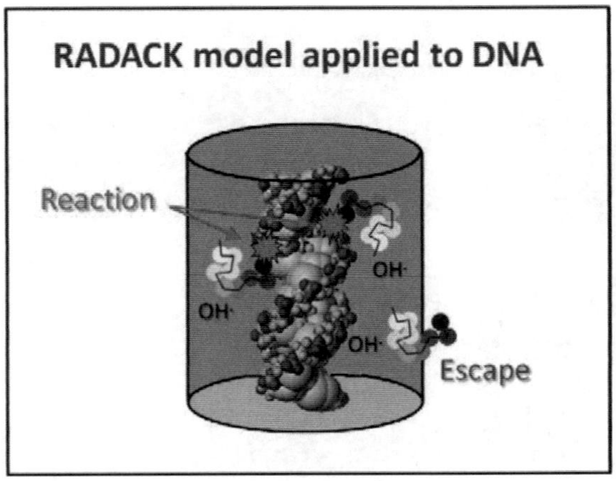

Figure 4. Principle of RADACK model. DNA with atoms represented either by van der Waals spheres (non-reactive atoms) or Smoluchowski's sphere (reactive atoms, spheres with radius proportional to their reactivity) in a box where OH· radicals are generated randomly. They can diffuse and either react with the encountered DNA atom or escape from the box. The number of encounters with the atoms of a nucleotide divided by the total number of generated OH· radicals represents the relative probability of reaction with that nucleotide.

2. Binding of small molecules affects DNA radiolysis

The use of the same combined electrophoresis and molecular modeling methodology allowed observing and understanding the protection or sensitization of certain regions of DNA by small positively charged molecules such as polyamines (putrescine, spermidine and spermine) [11-12] or some thiols (metabolites of the chemo- and radio-protective drug Ethyol (Amifostine) and related compounds, cysteamine, N-acetycysteine, captopril) [13-16].

As revealed by molecular modeling calculations, the regions of DNA that have a narrow minor groove and thus a strongly negative electrostatic potential attract such molecules. Through binding in the minor groove of DNA, they mask and therefore protect the H4' and H5'2 atoms whose abstraction would lead to strand breakage. When the most negative potential is located in the major groove, binding of the small ligands to such regions do not protect the critical sugar H atoms and therefore these regions are not protected. Moreover, the binding in the major groove can trigger in some cases the broadening of the minor groove and thus, can lead to the radiosensitization of the region (figure 5).

Thus the effects of polyamines and thiols are also dependent on the sequence of DNA via the modulation of DNA structure and electrostatic potential. It can be explained by the physical screening of the binding site, the scavenging of OH· radicals, the chemical repair of DNA radicals (only for thiols who donate the H atom of their SH function and thus repair the damaged sugar) and by the binding-induced structural changes of DNA (narrowing or broadening of the minor groove, or even compaction of DNA in the case of highly charged polyamines).

3. Binding of proteins affects DNA radiolysis

The DNA ligands of the highest biological interest are proteins. Their binding to DNA influences many critical cellular processes such as regulation of gene expression, DNA repair or DNA compaction. We have observed the protection of the binding sites of proteins such as lactose repressor (a protein managing the negative regulation of genes involved in lactose metabolism of *E. coli*), CRP (the catabolite gene activator protein, a regulatory protein involved in the energy metabolism of *E. coli*), an estrogen receptor (a ligand-activated transcription factor regulating the expression of genes controlled by estrogens), MC1 (a chromosomal protein of the archaebacterium Methanosarcina sp. involved in DNA bending and supercoiling) and histones (the octamer of proteins located in the center of nucleosome cores in chromatin) (figure 6). On the basis of this observation we have proposed the radiolytic footprinting as a useful method of determination of proteins binding sites on DNA [2].

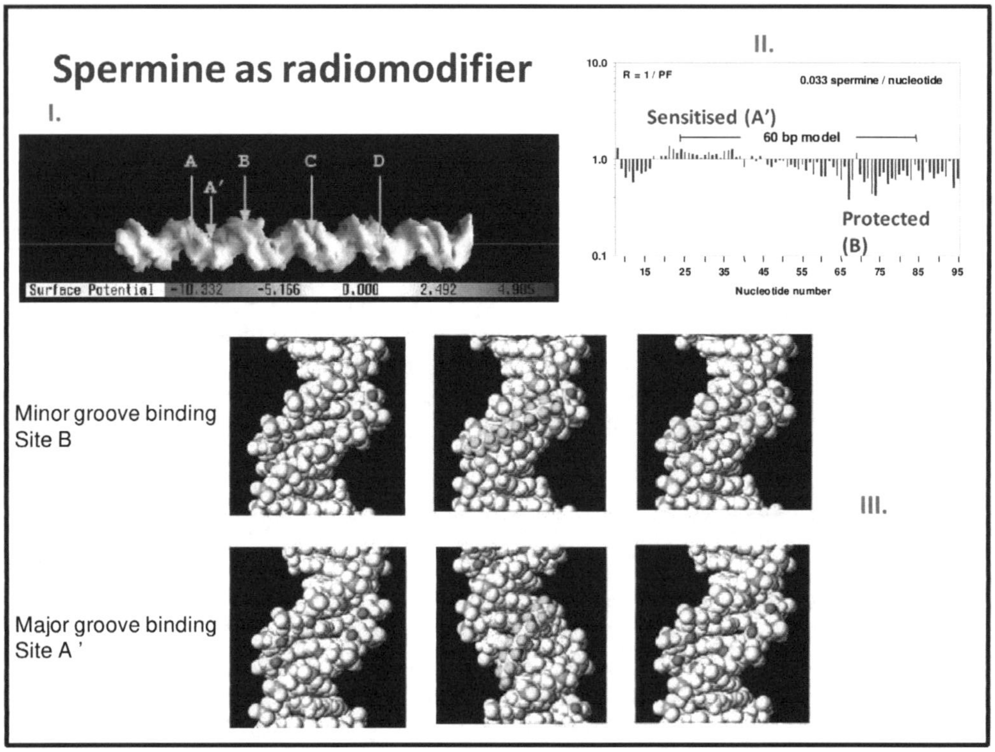

Figure 5. Radioprotection and radiosensitization of different regions of a DNA fragment by spermine (electrical charge +3). I. Electrostatic potential at the surface of DNA: the sites with a highly negative potential are red. A, B, C and D sites are located in the minor groove, and A' site is in the major groocve. II. experimentally determined values of 1/PF where PF (protection factor) = FSB in DNA irradiated in presence of spermine/FSB in free DNA). III. Molecular modeling showing the narrowing of the minor groove resulting from spermine binding at B site (minor groove) and broadening of the minor groove resulting from the binding of spermine at A' (major grove).

RADACK calculations in which the three dimensional structures of the complexes were extracted from the PDB structural data base were performed (Figure 7). They were in good agreement with the experimental results and allowed showing that the protection by the proteins is due to the physical screening of the binding site, to the scavenging of $OH^{\cdot}$ radicals by the protein and to the binding-induced structural changes of DNA [17-22].

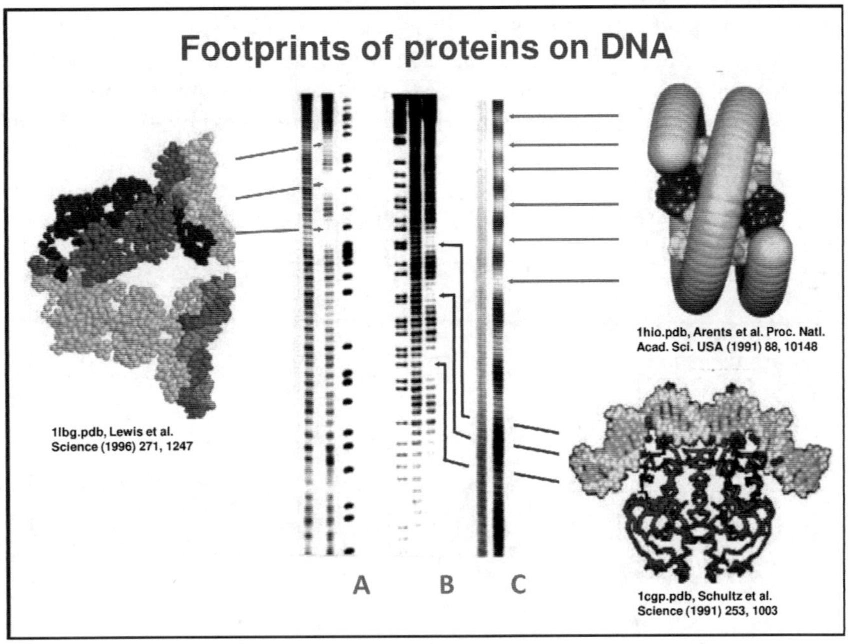

Figure 6. Variation of strand breaks frequency along a DNA irradiated in complex with lactose repressor (A), CRP (B) and with histone octamer in nucleosome core (C). The bleached zones correspond to the regions with low FSB yields due to the presence of the protein.

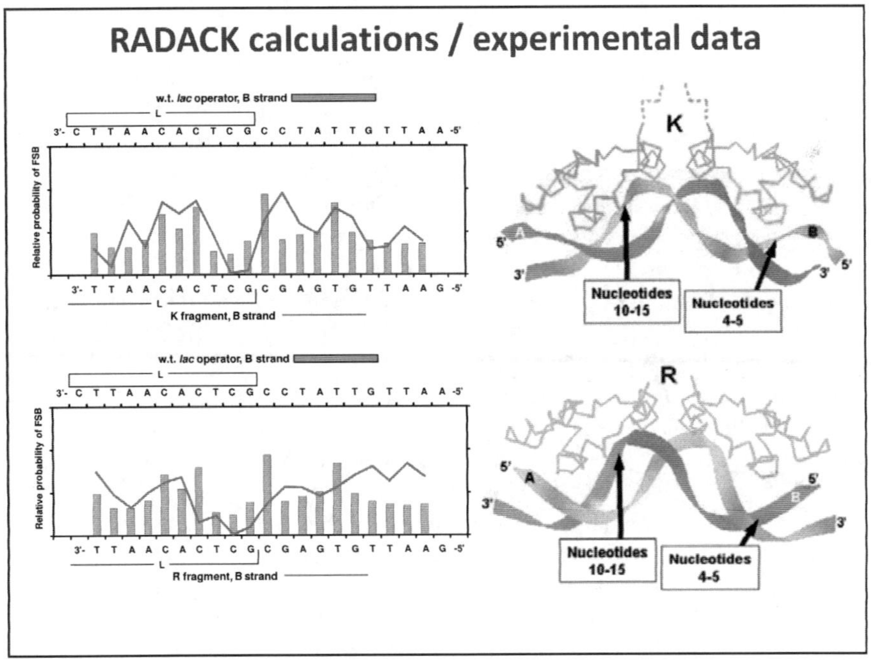

Figure 7. Comparison of RADACK calculations (red curves) with experimental results (FSB, histograms) for DNA irradiated in complex with the repressor headpieces. The three dimensional structures of the complex used in the calculations are 1LBG (K) and 1CJG (R).

4. Protein radiolysis destabilizes DNA-protein complexes

The protein cannot accomplish anymore its role of DNA "bodyguard" when irradiated with high doses of radiation because in reacting with the OH· radicals, it gets much damage and consequently, loses its ability to bind DNA. The type of reactions and consequences are quite similar to those of DNA: abstraction of H atoms and binding to aromatic rings, leading to backbone breakage and modification of side chains (figure 8).

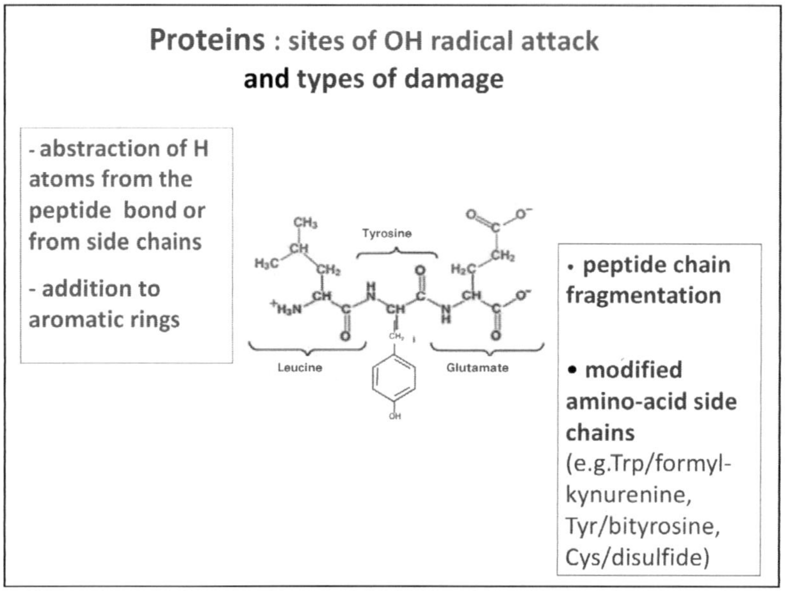

Figure 8. Attack of proteins by the OH· radicals produced by the radiolysis of water and its consequences.

Side chains oxidation of amino-acids are induced not only by radiation, but also by natural metabolic processes. These last ones are leading to deleterious effects when exceeding the detoxification ability of the organism (like in aging or strong oxidative stress conditions).

When irradiating DNA-protein complexes with high doses, the lactose repressor-operator complex, the complex between the formamidopyrimidine-DNA glycosylase (Fpg) and DNA bearing a model oxidative damage and the complex between estrogen receptor and the estrogen receptor element were destabilized [23-24]. In all cases we have proven that the loss of binding ability of the protein is the main cause of the complex destabilization. The results of a molecular modeling study in which 8-oxoG lesions were introduced into DNA were consistent with only a weak effect of DNA damage on the lactose repressor-operator interactions [25].

Moreover we have shown that when irradiated in complex with DNA, the protein loses its binding ability at doses higher than those necessary for inactivating the free protein. This means that when irradiated in the complex, the protein is less damaged than when irradiated free in solution. We have concluded that in the complex not only the protein is protecting DNA, but in turns, DNA protects the protein by masking (and thus rendering inaccessible for the OH· radicals) the aminoacids of the binding site. The results for the lactose repressor-operator complex are shown in figure 9.

RADACK calculations have predicted which amino-acids are most likely to be damaged by irradiation. Consistent with these predictions, mass spectrometry and fluorescence measurements have shown that the tyrosine residues are oxidized upon irradiation of the DNA binding domains (headpieces) of the lactose repressor [26]. Molecular dynamics simulation have shown that the replacement of tyrosine residues by their oxidation product 3,4-dihydroxyphenylalanine (DOPA) in the headpiece leads to a loss of its stability and to a reduction of its helical content of the protein.

Journal of Physics: Conference Series **261** (2011) 012010 doi:10.1088/1742-6596/261/1/012010

These results are in good agreement with the unfolding and destabilization of the headpiece as revealed by circular dichroïsm measurements [27].

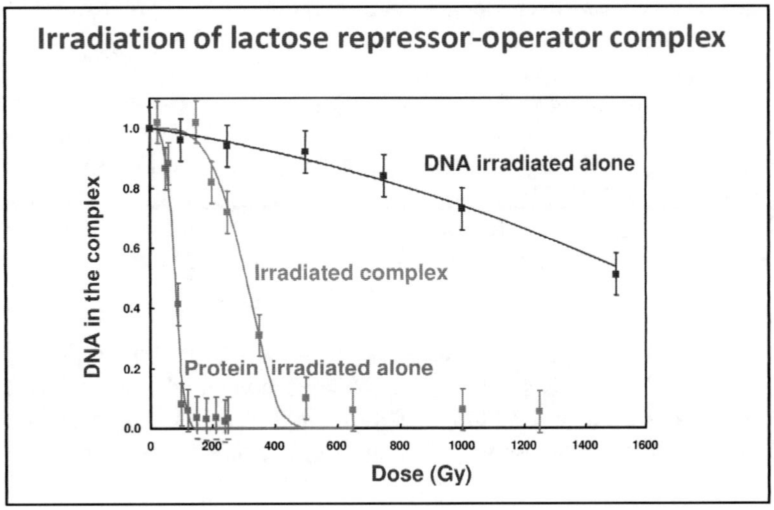

Figure 9. Decrease of DNA amount in DNA-repressor complexes upon the irradiation of the complex (magenta), of the free protein (red) or of the free DNA (blue).

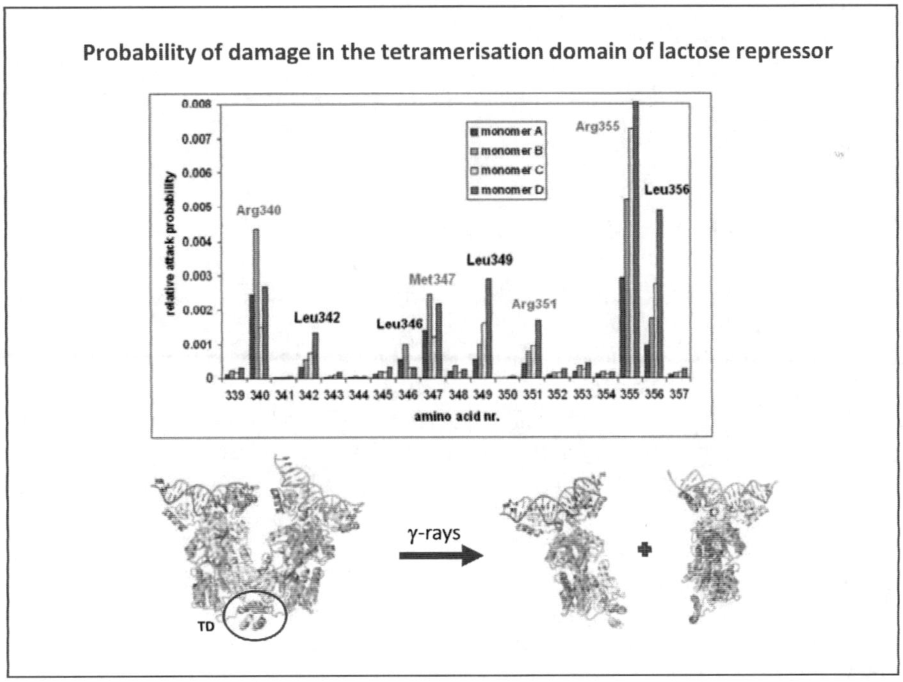

Figure 10. Probability of reaction of OH· radicals with the amino-acids of the tetramerization domain of the *lac* repressor tetramer. In the bottom a schematic representation of the tetramer-to-dimer radio-induced transition.

After comparing a native and an oxidized headpiece, molecular dynamics simulations were performed for 1. the native complex formed by two headpieces and the operator DNA, and 2. the complex between the oxidized headpiece and DNA. The results have shown effects consistent with complex destabilization: increased flexibility, increased DNA bending, modification of the hydrogen bond network, decrease of the positive electrostatic potential at the protein surface and of the global energy of DNA-protein interactions [28].

Another modification of the repressor structure resulting from the irradiation concerns its oligomerization state. Lactose repressor is a tetrameric protein formed by two identical homodimers connected by a 4-helix bundle based on leucine zippers (called tetramerization domain). We have shown that upon high dose irradiation the tetramer splits into its constitutive dimers likely because of the damage of the leucines of the tetramerisation domain. RADACK calculations (figure 10) have shown that these leucines whose integrity is indispensable for the maintenance of the tetramerization have indeed a high probability of getting damaged by $OH^{\cdot}$ radicals [29].

5. Conclusion

Our work allowed to observe and to explain the variations of radiation-induced damage yields and distribution in free DNA and in DNA involved in supramolecular complexes. It emphasized the critical role of structural parameters in the induction of damage. It also shows that the observed destabilization of DNA-protein interactions is mainly due to the chemical and structural damages of the protein.

Acknowledgments

All members of our two teams having contributed to the research on radiation damage to biomolecules as well as our co-workers are warmly thanked. Our research got the financial support of CNRS, INSERM, Electricity of France, League against Cancer, Association for Research against Cancer, Academy of Sciences of the Czech Republic (B1048901, A1048103, KJB4048401), MEYS CR (1P05OC085, OC09012), bilateral French-Czech programs (CNRS-ASCR, Barrande-Egide) and the European actions COST P9 and CM0603.

References

[1] Spotheim-Maurizot M, Charlier M and Sabattier R 1990 *Int. J. Radiat. Biol.* **57** 301
[2] Franchet-Beuzit J, Spotheim-Maurizot M, Sabattier R, Blazy-Baudras B and Charlier M 1993 *Biochem.* **32** 2104.
[3] Sy D, Savoye C, Běgusová M, Michalik V, Charlier M and Spotheim-Maurizot M 1997 *Int. J. Radiat. Biol.* **72** 147.
[4] Běgusová M, Spotheim-Maurizot M, Sy D, Michalik V and Charlier M 2001 *J. Biomol. Struct. Dyn.* **19** 141.
[5] Tartier L, Michalik V, Spotheim-Maurizot M, Rahmouni A R, Sabattier R and Charlier M 1994 *Nucleic Acids Res.* **22** 5565.
[6] Michalik V, Spotheim-Maurizot M and Charlier M 1995 *J. Biomol. Struct. Dyn.* **13** 565.
[7] Tartier L, Spotheim-Maurizot M and Charlier M 1998 Int. J. Radiat. Biol. **73** 45.
[8] Běgusová M, Tartier L, Sy D, Michalik V, Spotheim-Maurizot M and Charlier M 1999 *Int. J. Radiat. Biol.* **75** 913.
[9] Nikjoo H, O'Neill P, Terrissol M and Goodhead DT 1999 *Radiat. Environ. Biophys.* **38** 31.
[10] Friedland W, Paretzke HG, Ballarini F, Ottolenghi A, Kreth G and Cremer C 2008 *Radiat. Environ. Biophys.* **47** 49.
[11] Spotheim-Maurizot M, Ruiz S, Sabattier R and Charlier M 1995 *Int. J. Radiat. Biol.* 68 571.
[12] Sy D, Hugot S, Savoye C, Ruiz S, Charlier M and Spotheim-Maurizot M 1999 *Int. J. Radiat. Biol.* **75** 953.
[13] Spotheim-Maurizot M, Franchet J, Sabattier R and Charlier M 1991 *Int. J. Radiat. Biol.* **59** 1313.

Journal of Physics: Conference Series **261** (2011) 012010 doi:10.1088/1742-6596/261/1/012010

[14] Spotheim-Maurizot M, Garnier F, Kieda C, Sabattier R and Charlier M 1993 *Radiat. Environ. Biophys.* **32** 337.
[15] Savoye C, Swenberg C, Hugot S, Sy D, Sabattier R, Charlier M and Spotheim-Maurizot M 1997 *Int. J. Radiat. Biol.* **71** 193.
[16] Sy D, Durand C, Hugot S, Savoye C, Swenberg C, Charlier M and Spotheim-Maurizot M 1999 *Theor. Chem. Acc.* **101** 114.
[17] Isabelle V, Franchet-Beuzit J, Sabattier R, Laine B, Spotheim-Maurizot M and Charlier M 1993 *Int. J. Radiat. Biol.* **63** 749.
[18] Běgusová M, Sy D, Charlier M and Spotheim-Maurizot M 2000 *Int. J. Radiat. Biol.* **76** 1063.
[10] Běgusová M, Eon S, Sy D, Culard F, Charlier M and Spotheim-Maurizot M 2001 *Int. J. Radiat. Biol.* **77** 645.
[20] Běgusová M, Giliberto S, Gras J, Sy D, Charlier M and Spotheim-Maurizot M 2003 *Int. J. Radiat. Biol.* **79** 385.
[21] Davídková M, Štísová V, Goffinont S, Gillard N, Castaing B and Spotheim-Maurizot M 2006 *Radiat. Prot. Dosim.* **122** 100.
[22] Štísová V, Goffinont S, Spotheim-Maurizot M and Davídková M 2006 *Radiat. Prot. Dosim.* **122** 106.
[23] Eon S, Culard F, Sy D, Charlier M and Spotheim-Maurizot M 2001 *Radiat. Res.* **156** 110.
[24] Gillard N, Běgusová M, Castaing B and Spotheim-Maurizot M 2004 *Radiat. Res.* **162** 566.
[25] Sy D, Flouzat C, Eon S, Charlier M and Spotheim-Maurizot M 2001 *Theor. Chem. Acc.* **106** 137.
[26] Gillard N, Goffinont S, Buré C, Davídková M, Maurizot J C, Cadene M and Spotheim-Maurizot M 2007 *Biochem. J.* **403** 463.
[27] Mazier S, Villette S, Goffinont S, Renouard S, Maurizot J C, Genest D and Spotheim-Maurizot M 2008 *Radiat. Res.* **170** 604.
[28] Aci-Sèche S, Garnier N, Goffinont S, Genest D, Spotheim-Maurizot M and Genest M 2010 *Eur. Biophys. J.* **39** 1375.
[29] Goffinont S, Davídková M and Spotheim-Maurizot M 2009 *Biochem. Biophys. Res. Commun.* **386** 300.

Effects of reactive oxygen species on metabolism monitored by longitudinal ^{1}H single voxel MRS follow-up in patients with mitochondrial disease or cerebral tumors

J M Constans[1,2], S Collet[1,2], J S, Guillamo[2,7], G Hossu[1,9], S Lacombe[10], Y A Gauduel[12], C Houée Levin[10], W Dou[4], S Ruan[5], L Barré[13], F Rioult[6], J M Derlon[7], E Lechapt-Zalcman[2,8], S Valable[2], F Chapon[8], P Courtheoux[1], V Fong[11] and F Kauffmann[3]

[1]MRI Unit, Caen University Hospital, Caen, Normandy, France - [2]CERVOxy Group, CI-NAPS, UMR 6232 CI-NAPS , Cyceron, Caen, Normandy, France - [3]Mathematics LMNO CNRS UMR 6139, Caen University, Caen, Normandy, France - [4]Tsinghua University, Beijing, China - [5]CReSTIC EA 3804, IUT Troyes, Troyes, France - [6]CNRS UMR 6072, GREYC, Caen, Normandy, France - [7]Neurosurgery and Neurology, Caen University Hospital, Caen, Normandy, France - [8]Pathology, Caen University Hospital, Caen, Normandy, France - [9] CICIT INSERM, Nancy University, Nancy,France - [10] Paris-Sud 11 University-CNRS, Orsay, France - [11] Caen University, France - [12] LOA, Ecole Polytechnique - ENSTA ParisTech, Palaiseau, France - [13] GDMTEP, Group CI-NAPS, UMR 6232 CI-NAPS, Cyceron, Caen, France.

E-mail: constans-jm@chu-caen.fr

Abstract. Free radicals, or Reactive Oxygen Species (ROS), have an effect on energy and glycolytic metabolism, mitochondrial function, lipid metabolism, necrosis and apoptosis, cell proliferation, and infiltration. These changes could be monitored longitudinally (every 4 months over 6 years) in humans with glial brain tumors (low and high grade) after therapy, using conventional magnetic resonance imaging (MRI) and spectroscopy (MRS) and MR perfusion. Some examples of early clinical data from longitudinal follow-up monitoring in humans of energy and glycolytic metabolism, lipid metabolism, necrosis, proliferation, and infiltration measured by conventional MRI, MRS and perfusion, and positron emission tomography (PET) are shown in glial brain tumors after therapy. Despite the difficulty, the variability and unknown factors, these repeated measurements give us a better insight into the nature of the different processes, tumor progression and therapeutic response.

1. Introduction

Diagnosis of cancer and mitochondrial diseases is challenging despite the availability of large numbers of imaging and biochemical investigational techniques and data. The conventional magnetic resonance imaging (MRI) provides information on tumor anatomy, its extent, and pathology; however the specificity of diagnosis distinguishing between benign and malignant disease is still not optimal. Furthermore, MRI does not provide information on the underlying biochemical processes associated with tumor progression and regression. Thus, the biochemical information from tissues may be useful for specific diagnosis and accurate prognosis when used as an adjunct to anatomic details obtained from MR images. This can be achieved through in vivo Magnetic Resonance Spectroscopy

(MRS) that provides unique metabolic information, at the molecular level, of diagnostic and prognostic importance for tumor typing, and grading. The tissues contain several biologically important metabolites (biochemicals) in addition to water and fat hydrogen atoms that could be followed by MRS. It also provides information on the alterations in metabolic pathways during disease processes by detection and quantification of metabolites. MRS is also a useful tool in the planning and evaluation of treatments and in predicting tumor progression and treatment response. It is known that tumor growth is not a simple regulated phenomenon. There are variations in metabolism and cell density, and in the presence of cystic and necrotic regions (particularly in malignant tumors). Thus, the MR spectrum obtained from a tumor tissue can be a mixture of different tissue types, including those of different grades, which complicates the spectral interpretation. Most MRS measurements are based on the use of proton (^{1}H) nuclei, thanks to its great sensitivity. The integration of ^{1}H MRS with clinical MRI investigation is relatively straightforward and easy to implement. Addition of in vivo ^{1}H MRS protocol increases the overall acquisition time by approximately 10 to 20 minutes but it drastically improves the diagnostic and prognostic accuracy. This method has generated considerable interest in recent years and the added value of the *in vivo* ^{1}H MRS is in process of evaluation in clinical application. In particular it would allow quantification of metabolites from a well-defined volume of interest (VOI) or volume element (voxel). Moreover, the heterogeneous nature of tumors could be understood and detected, which would be of great interest for biopsy guidance and planning, monitoring, and evaluation of treatments such as surgery, radiotherapy, and chemotherapy. Therefore, the possibility to obtain *in vivo* quantitative biochemical information has revolutionized the field of clinical neurooncology. In the present work, metabolic information obtained in MRS measurements, associated with anatomic data, allows us to establish an accurate prognosis based on a spectral and metabolic classification system. In addition, in longitudinal follow-up studies, multimodality (with Positon Emission Tomography (PET) and perfusion) allow us to study the relationship between MRS, PET data, MRI segmentation and perfusion, together with some biologic parameters (eg, proliferation, necrosis, lactate, glutamine, some Reactive Oxygen Species (ROS) and their consequences, and anti-oxidants) thus helping to characterize sensitive biomarkers and to detect early changes in order to assess rapidly and more accuratly therapies. Moreover, MRS data to be evaluated have to be related to data such as the response to therapy, histology, and genetic prognostic mutations (Isocitrate Dehydrogenase 1 IDH1, 1p19q, Methyle guanine methyl transferase MGMT).

1.1 Contribution of MRS in management of brain tumors

Numerous studies reported in the literature have shown that ^{1}H MRS contributes significantly to the management of brain tumors. The major metabolites seen at long, intermediate and short echo times are N-Acetyl-aspartate (NAA; 2.02 ppm), total creatine (tCr; 3.03 ppm) and choline containing compounds (tCho; 3.2 ppm). NAA is found predominantly in neurons. The tCr level is reduced in astrocytomas and is nearly absent in meningiomas and schwannomas, indicating changes in energy metabolism. Gliomatosis cerebri is an uncommon and infiltrative type of glioma which is difficult to diagnose but has higher levels of tCr, thus distinguishing it from low-grade gliomas [1]. The tCho/tCr ratio increases with the grade of the glioma [2] but the large overlap of ratios between grades prevents its individual use as a reliable grading index. The level of tCho increases from grade II to grade III in astrocytomas, but grade IV tumors show wide variation due to the heterogeneity in cell density and necrosis. In low-grade pediatric gliomas, citrate was first found to be elevated [3] followed by an increase in tCho as the tumor progressed. The ratio tCho/NAA was shown to correlate with long-term survival independent of tumor type [4]. In addition to the above predominant metabolites observed at long echo times, additional metabolites such as Glx (glutamate (Glu) and glutamine (Gln), at 3.6-3.9 ppm and 2.1- 2.6 ppm) and myo-inositol (mI, at 3.56 ppm) are seen at shorter echo times and these also have diagnostic and prognostic potential. Myo-inositol is a major metabolite found in glia but not in neurons [5] and has a role as an osmolyte. Elevated mI is seen in low-grade gliomas, especially in oligodendrogliomas, however it is decreased in high-grades. Higher Glx and glutathione (GSH, multiplet at 3.8 ppm) are seen in meningiomas compared to astrocytomas [6]. Alanine (Ala, doublet at

1.47 ppm) is also seen, which is very suggestive of meningioma [7]. These observations may indicate that energy metabolism is mainly carried out by partial oxidation of glutamine rather than glycolysis, with alanine being the end product instead of lactate. Elevated Gln is also found in some oligodendrogliomas, which could help (in conjunction with high mI) discriminate them from astrocytomas [8]. Lactate (a doublet at 1.33 ppm) is observed in tumours resulting from glycolysis [9], and lipids (1.3 and 0.9 ppm) are seen in high-grade tumors [10;11;12].

1.2 ROS and mitochondrial tumor suppressor genes roles in tumor development

The metabolic adaptation in tumors, the Warburg effect, induces a high rate of glycolysis in tumors, and sometimes a concurrent defect in mitochondrial respiration. These changes are hypothesized to be due to mitochondrial tumor suppressor genes. For example, succinate dehydrogenase and fumarate hydratase [13] are mitochondrial proteins of the tricarboxylic acid (TCA) cycle and the respiratory chain, and when mutated lead to tumors (for succinate dehydrogenase); it is known as paragangliomas and pheochromocytomas. Recently, a novel mitochondrial protein, SDHAF2 (SDH5), has also been shown to be a paraganglioma-related tumor suppressor gene. Another mitochondrial and TCA cycle related protein, isocitrate dehydrogenase 1(IDH1), uses Nicotinamide adénine dinucléotide phosphate hydrated (NADPH) as a cofactor and is frequently mutated in glioblastomas [14]. It should be noticed that an accumulation of 2-hydroxyglutarate (2-HG) is associated with this IDH1 mutation [15]. There are currently many competing hypotheses concerning the role of these genes in tumorigenesis, but frequently recurring themes are the stabilization of hypoxia-inducible factor 1-alpha (HIF1-alpha) and upregulation of genes involved in angiogenesis, glucose transport, and glycolysis. Other postulated mechanisms include the inhibition of developmental apoptosis, altered gene expression due to histone deregulation, and the acquisition of novel catalytic properties. Effects of these gene mutations on metabolism are not clear, but they can be studied by MRS. It has been suggested in particular that oxidative stress and thus ROS formation could be actively involved in these processes (cf. figure 1). ROS play a role in the regulation of cell survival or death by mediating the apoptosis, but surprisingly ROS can also favor an opposite effect such as the tumor cell proliferation. Tumor cells manage to control high extracellular ROS production and this prevent apoptosis induction by ROS but still maintain a sufficient level of ROS for proliferation stimulation. [16]

1.3 Studying the relationship between the MRS measures, ROS and antioxidants

Detection of the metabolites and processes using MRS is crucial to study oxidative stress and redox balance. We suggest that using MRS to identify metabolites (especially those from the Krebs cycle seen in blue in figure 1) involved in the formation of ROS and glutathione (GSH), could inform about the balance between ROS and antioxidant formation involved in some diseases and tumor processes and thus helps to better understand the role of ROS particularly in cancer [17,18,19].

It is known that succinate oxidation can support the higher rate of ROS production in non-phosphorylating mitochondria. Accumulation of succinate [13] during the hypoxic period might also contribute to increase mitochondrial ROS generation upon reoxygenation. The quantification of succinate and acetate *in vivo* can be correlated with the ROS production. However, *in vivo* succinate comes also from bacteria or from impaired tissue respiratory chain. Similarly, ROS could come from impaired tissue respiratory chain, from macrophages or from neutrophils.

MRS enables us to characterize pathologic processes such as proliferative potential, which correlates with a high level of choline, particularly in gliomas. These data could make possible the study of processes put into place by the cells in response to oxidative stress and the involvement of ROS in the proliferation of tumor cells, in particular by activation of HIF and angiogenesis [20].

Necrosis can also be characterized in MRS by the quantification of free phospholipids. This knowledge is useful in the study of lipid peroxidation and necrosis/apoptosis induced by ROS and therapies. Lactate is observed in tumors due to the metabolic switch that occurs when the Krebs cycle is altered and glycolysis increases. It is known that ROS are involved in these changes through the regulation of glycolysis enzymes, and studying lactate by MRS is thus used to better understand this ROS-induced metabolic switch. A correlation was not always found between lactate levels and tumor grade or metabolic rate. Lactate accumulation occurs in both intra- and extra-cellular spaces and its overall level is a function of metabolic rate and clearance; hence in higher-grade tumors lactate may pool in necrotic or cystic regions independent of increased glycolysis. In addition, this type of study gives some insight into pH change in tumors, which is important for therapeutic efficacy.

It would be of interest to do non invasive longitudinal follow-up studies of these metabolites as well as pathological processes such as proliferation, hyperperfusion, necrosis, ROS, and disease processes.

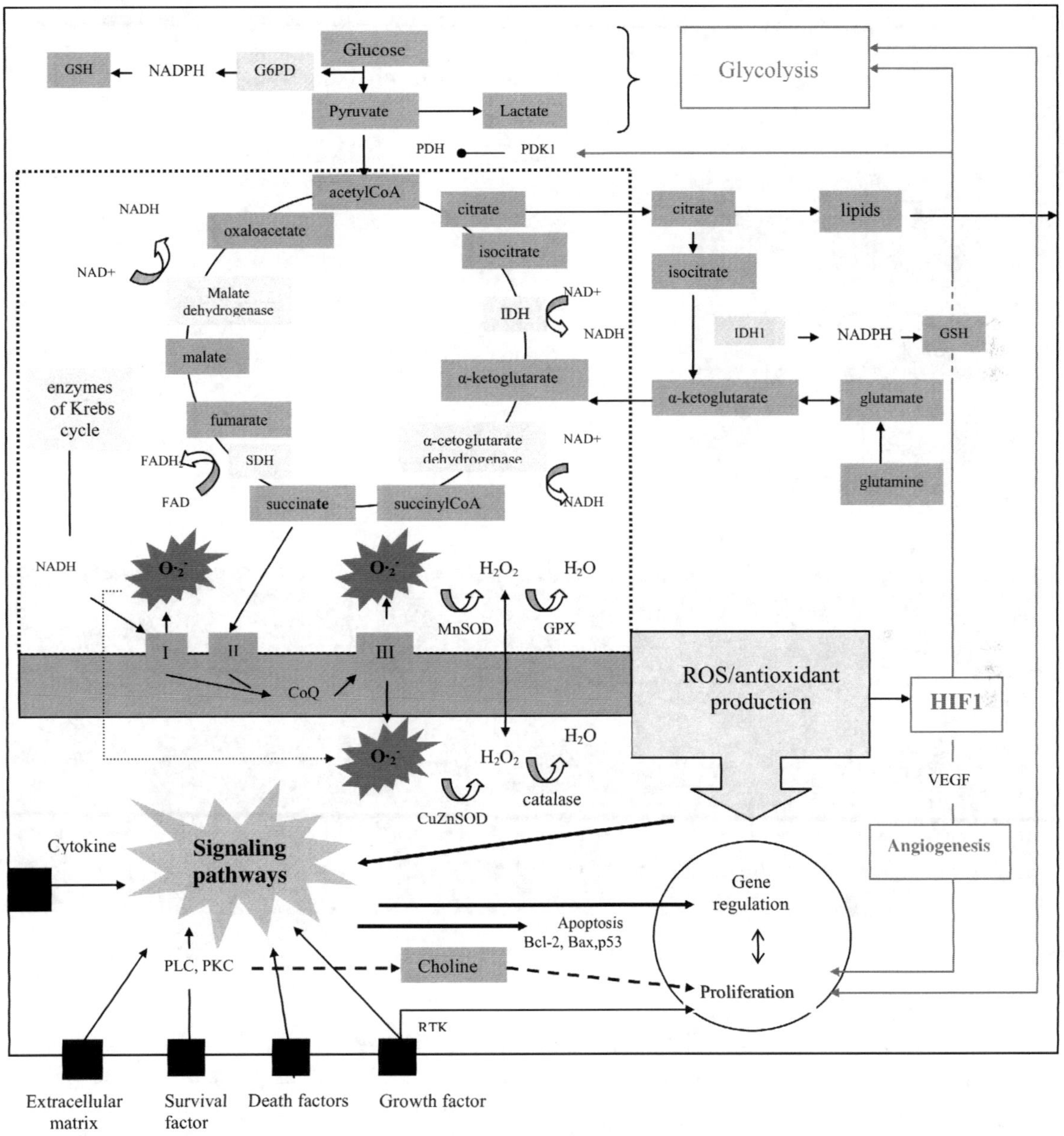

Figure 1. Metabolites (measurable by MRS, in blue) and metabolism involved in ROS and antioxidant balance

Journal of Physics: Conference Series **261** (2011) 012011 doi:10.1088/1742-6596/261/1/012011

2. Examples of potential effects of ROS on the metabolism in different pathologies

MRS is a non-invasive and reproducible technique for monitoring this energy metabolism *in vivo* in normal and diseases states. In this study, the aim of this technique was to try to better understand the potential relationship between free radicals and metabolism on these processes in some diseases such as mitochondrial disease, abscesses or cancer.

2. 1 Mitochondrial-dependent diseases

Mitochondrial-dependent diseases are a group of disorders caused by dysfunctional mitochondria. In these diseases, MRS is a tool which allows us to better understand the metabolism changes such as lactate production and evolution (figure 2) and their relationships with genetic, biochemical and free radical abnormalities. However, these diseases present some difficulties for MRS study. In fact, abnormalities and processes are heterogeneous and multifocal. In addition, there is a fair amount of temporal variation.

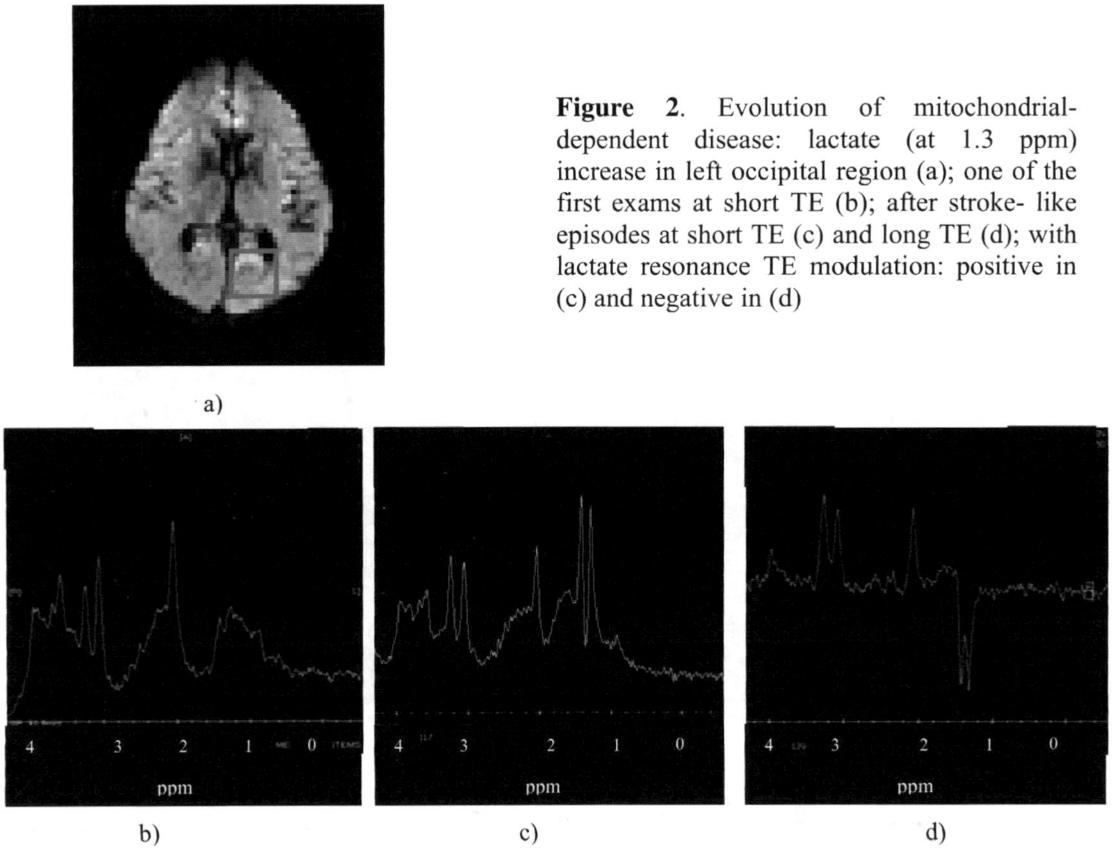

Figure 2. Evolution of mitochondrial-dependent disease: lactate (at 1.3 ppm) increase in left occipital region (a); one of the first exams at short TE (b); after stroke- like episodes at short TE (c) and long TE (d); with lactate resonance TE modulation: positive in (c) and negative in (d)

2.2 Gliomas (with and without treatment)

The World Health Organization (WHO) classification is based on the presumed cell origin, distinguishes astrocytic, oligodendrocytic, and mixed gliomas. A grading system is based on the presence of the following criteria: increased cellular density, nuclear atypia, mitosis, vascular proliferation, and necrosis. The main histological subtype of grade I gliomas are pilocytic astrocytomas, which are benign. Diffuse astrocytomas, oligodendrogliomas and oligoastrocytomas are low-grade (II) or high-grade (III and IV) tumors. Glioblastomas correspond to grade IV astrocytomas.

Characteristics of spectroscopic profiles of glioma often show an increase in Choline/N-Acetyl-Aspartate (Cho/NAA, choline at 3.22 ppm and NAA at 2.02 ppm), Cho/Cr linked to proliferation and Myoinositol/Creatine (mI/Cr) ratios, a decrease in NAA/Cr, correlate better with infiltration. Sometimes the spectral profile presents an increase in lactate and lipids which correlates with necrosis (figure 3).

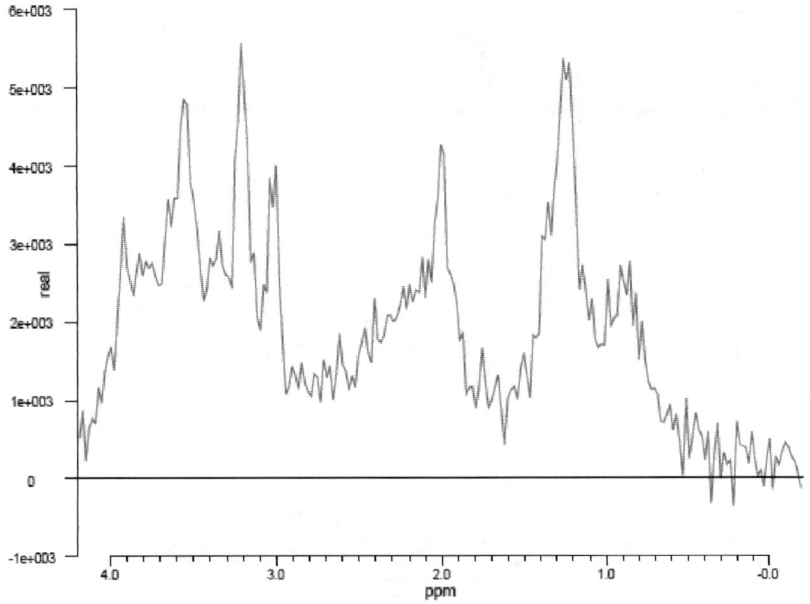

Figure 3. Example of spectral variation in glioma with SCI-MRS-LAB.
Cho (containing metabolites glycerophosphocholine (GPC) and phosphocholine (PC) at 3.22 ppm) elevated : proliferation. Lactate (at 1.33 ppm) : glycolysis. NAA/Cr (NAA at 2.02 ppm and Cr at 3.03 ppm) decreased: infiltration.

Cho/Cr and mI/ Cr decrease under chemotherapy (Temozolomide ®). NAA/Cr ratios are variable and have a tendency to improve under Temozolomide in gliomatosis. None or very little change in NAA/Cr occurs in oligodendrogliomas. A study of the contralateral side is very important after therapy to assess effects of these therapies on the remaining non- tumoral neuronal tissues. MRS can also contribute to the differential diagnosis between a tumoral and non tumoral lesion, especially in abscesses.

3. Methods and Results

3. 1 - MRI acquisition

Measurements of Sagittal T1, axial proton density, T2, FLAIR, diffusion, perfusion, 3D T1 and 3D T1 planes after gadolinium were acquired. Total tumoral volume, volume of FLAIR and T2 hypersignal, contrast enhancement, hyperperfusion and necrosis volumes are available on the MRI. A late enhancement should be taken into account. A segmentation allows a more quantitative multispectral MRI analysis to estimate tumor volumes, oedema and necrosis (figure 4, [21,22]). These data allow us to show that treated tumor volumes, observed on MRI, change little between two measurements while spectroscopic profiles and Cho/Cr or mI/Cr ratios decrease (see below section 3.5).

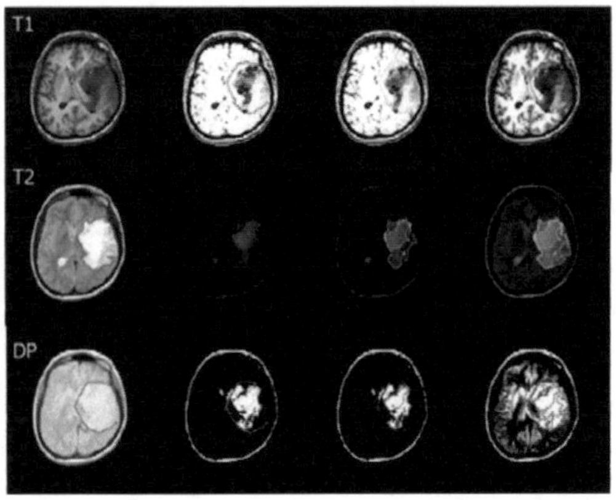

Figure 4. Manual segmentation and tissue classification of the different compartments from different weighted (T1, T2 and proton density) images from an oligodendroglioma

3.2 MRS acquisition, data processing and results

Single voxel (6 to 12 cm^3) from 1,5 T machine were done on most aggressive area (figure 5) and in contra-lateral side (figure 6) with PRESS MRS sequence and with multiple TE (35, 144, 288, 432ms). Measured volume size ranged from 1 cm^3 to 27 cm^3, Concentrations ranged from 0.25 mM to 100 mM, TE influence on the detected resonance number. The lowest statistical significant estimated concentration is in general around 1mM.

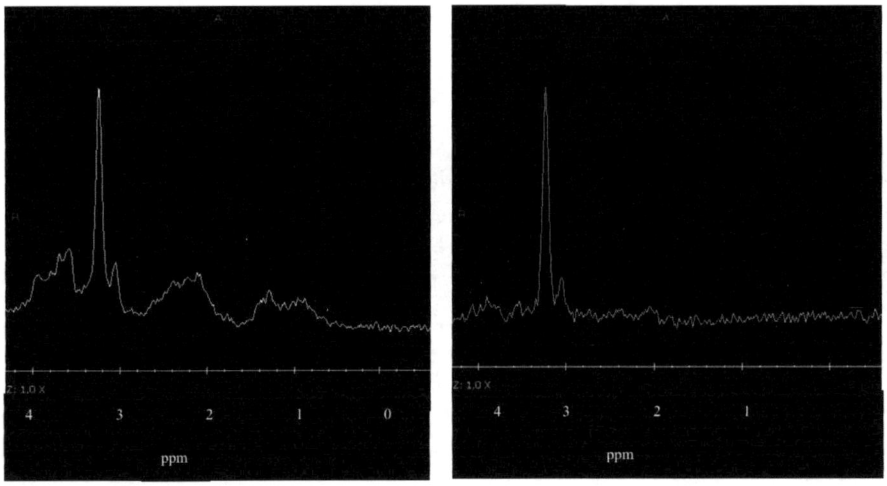

Figure 5. Example of spectra from glial tumor patients at 1.5 Tesla in PRESS sequences at different TE: 35 ms (left) and 144 ms (right)

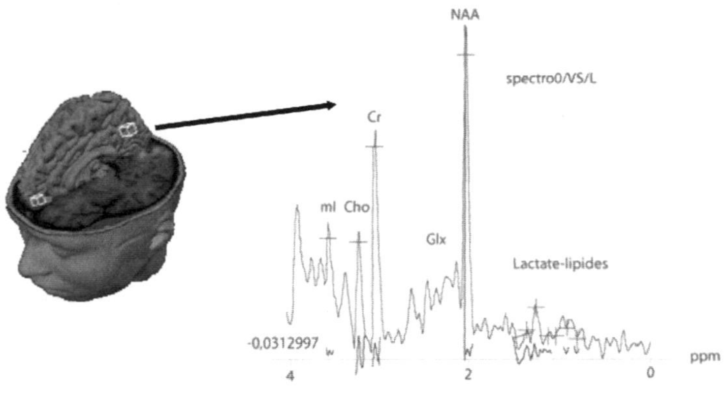

Figure 6. Magnetic Resonance Spectroscopy single voxel short TE 35 ms spectrum from white matter contra-lateral side close to a normal spectrum

MRS acquisition is sensitive to some parameters and some instrumental problems: localization quality, saturation bands and homogeneity [23]. This is even more evident in large volumes like CSI. SA/GE software (figure 7-a) and a home-written SCI-MRS-LAB [24,25] (at CHU and Caen University and in Scilab INRIA-ENPC open source code) automatic processing software yield amplitudes, areas, ratios, and relative concentrations. Quantification based on amplitude and area (proportional to concentration and relaxation) estimation in spectral domain needs solvent resonance(s) suppression and presents some difficulties such as unknown resonances and molecules metabolization. This process requires normalization with relative scale with ratio to another metabolite such as water, NAA, or Cr or absolute metabolite concentration (absolute scale).

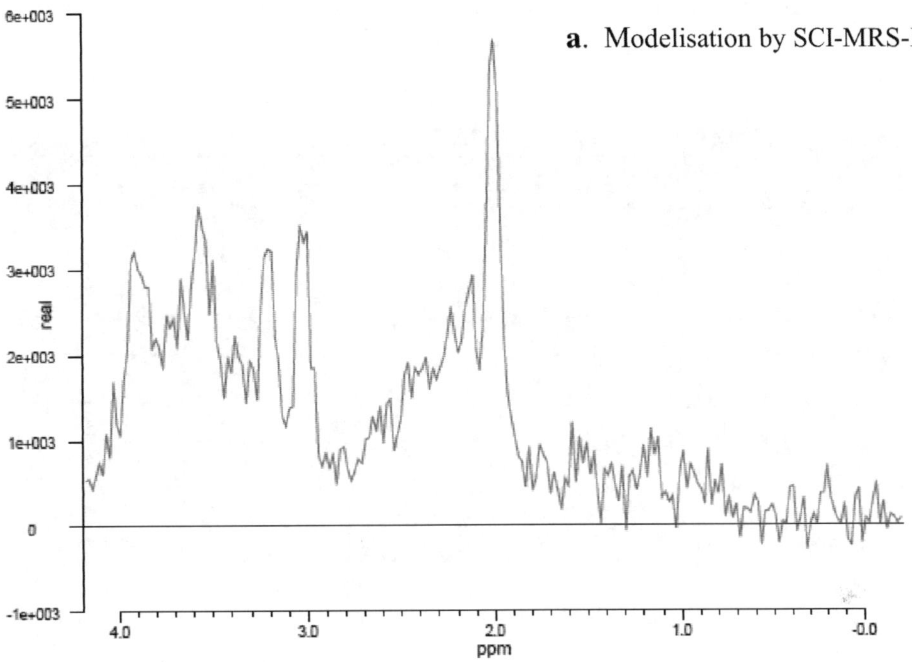

a. Modelisation by SCI-MRS-LAB

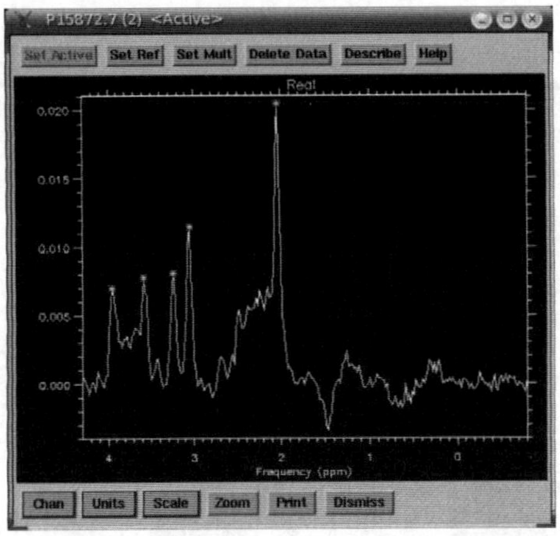

b. Spectral analysis software from General Electric (SA/GE)

Figure 7. Example and comparison of SCI-MRS-LAB (a) and SAGE (b) processing for short TE spectra from normal healthy volunteer

There are few quantification methods (SA/GE, JMRUI, SCI-MRS-LAB, sparse representation [30]), which include different processing steps like water suppression, baseline estimation, fitting and analysis [26-29]. Figure 7 shows an example of quantification methods comparison between Vendor tools for example SAGE (Spectroscopy Analysis by General Electric) (b) and SCI-MRS-LAB (a) Usually, we can obtain much better water signal suppression and baseline estimation with SCI-MRS-LAB than with some wavelets or SAGE. SCI-MRS-LAB gives also better results than SAGE for overlapped coupled spins as with glutamate and glutamine, citrate or an usual additional resonances to fit.

Acquisition of the spectral signatures in vitro of metabolites requires accurate protocol in order to identify and measure them (as for example GSH in figure 8) in patients. This protocol, optimized by Gabriela Hossu [29], optimizes the accuracy for frequency shift (eg, due to temperature or pH) of measurement and also the reproducibility. It is of great interest to be able to detect and measure directly or indirectly tissular antioxidant level with different therapies [34].

a **b** **c**

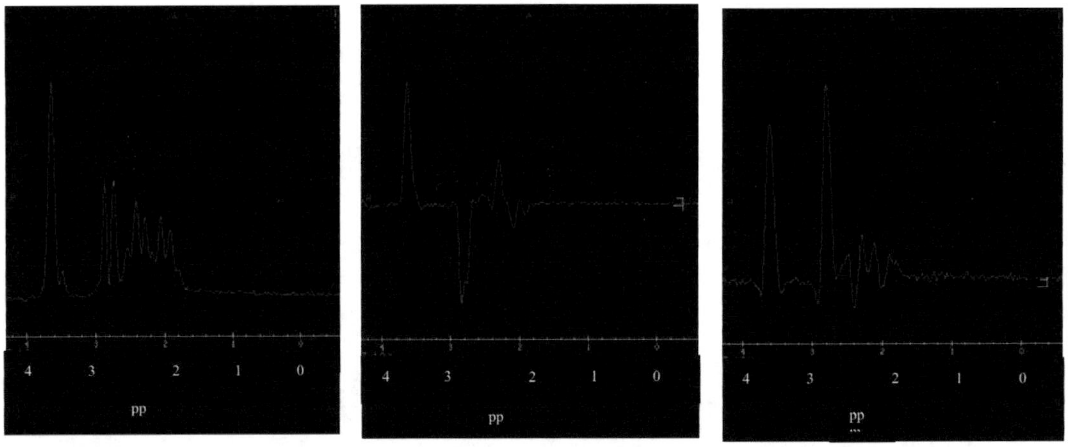

Figure 8. In vitro spectrum of glutathione (GSH): GSH resonances TE modulation: positive in a) at TE 35 ms, partially negative in b) at TE 144 ms and again positive at TE 288 ms.

3. 3 Perfusion

This NMR technique allows the study of cerebral microcirculation with a very fast slice acquisition after gadolinium injection (first past perfusion). The evaluation of new methods of perfusion in clinical conditions involves Spin labelling, first passage (figure 9), dynamic relaxometry, modelization of index size of vessels, evaluation of variability (reserve Vaso Constriction, Vaso Dilatation (Diamox, CO_2)).

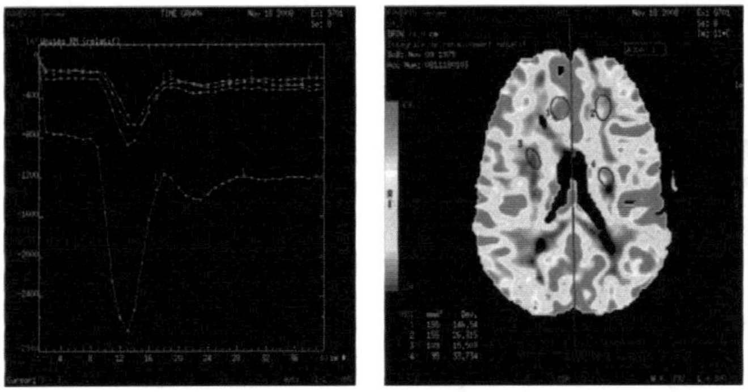

Figure 9. Glial tumor: Map (right) of relative (left) CBV ratio proportional to area under the curves to evaluate tumoral vascularization after intravenous gadolinium injection and showing hyperperfusion (in red) and hypoperfusion (in blue).

The data show some tumoral angiogenesis with important lesions of the BBB (blood-brain barrier). The extravasation of radiocontrast agents induces an increase in the curve of first passing above the baseline. The relative Cerebral Blood Volume (rCBV) in glial tumors is often elevated between 2 and 6 and shows hyperperfusion.

3.4 PET and multimodality

These techniques allow to study the relationship between MRI (FLAIR, T1, perfusion) and MRS. Two examples of relationships in the proliferation study and in the hypoxia and necrosis study with free radicals are given in figures 10 and 11 respectively.

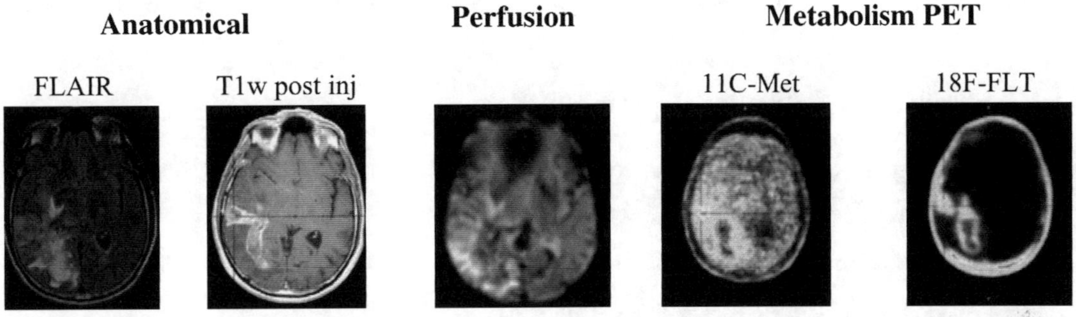

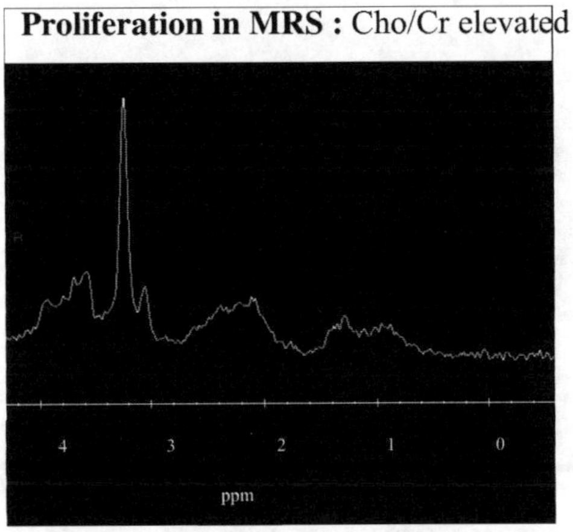

Figure 10. Proliferation study in PET and MRS (Cho/Cr): on this patient we can detect contrast enhancement on T1 post injection, hyperperfusion, hyperfixation on Met and FLT PET images and very important Cho/Cr increase on MRS (Cho at 3.2 ppm and Cr at 3.03 ppm) showing proliferation.

There is a large variability, but the repetition and the modelization of spectroscopic measurements during longitudinal follow-up allowed us to improve the prognostic evaluation. Studying the relationship between MRS measures, methionine and FLT PET quantitation, segmentation and perfusion parameters could lead to a better understanding of the therapeutic response, especially with regard to chemotherapy, antiangiogenic molecules, and, in the future, hypoxia modulators and other new therapies.

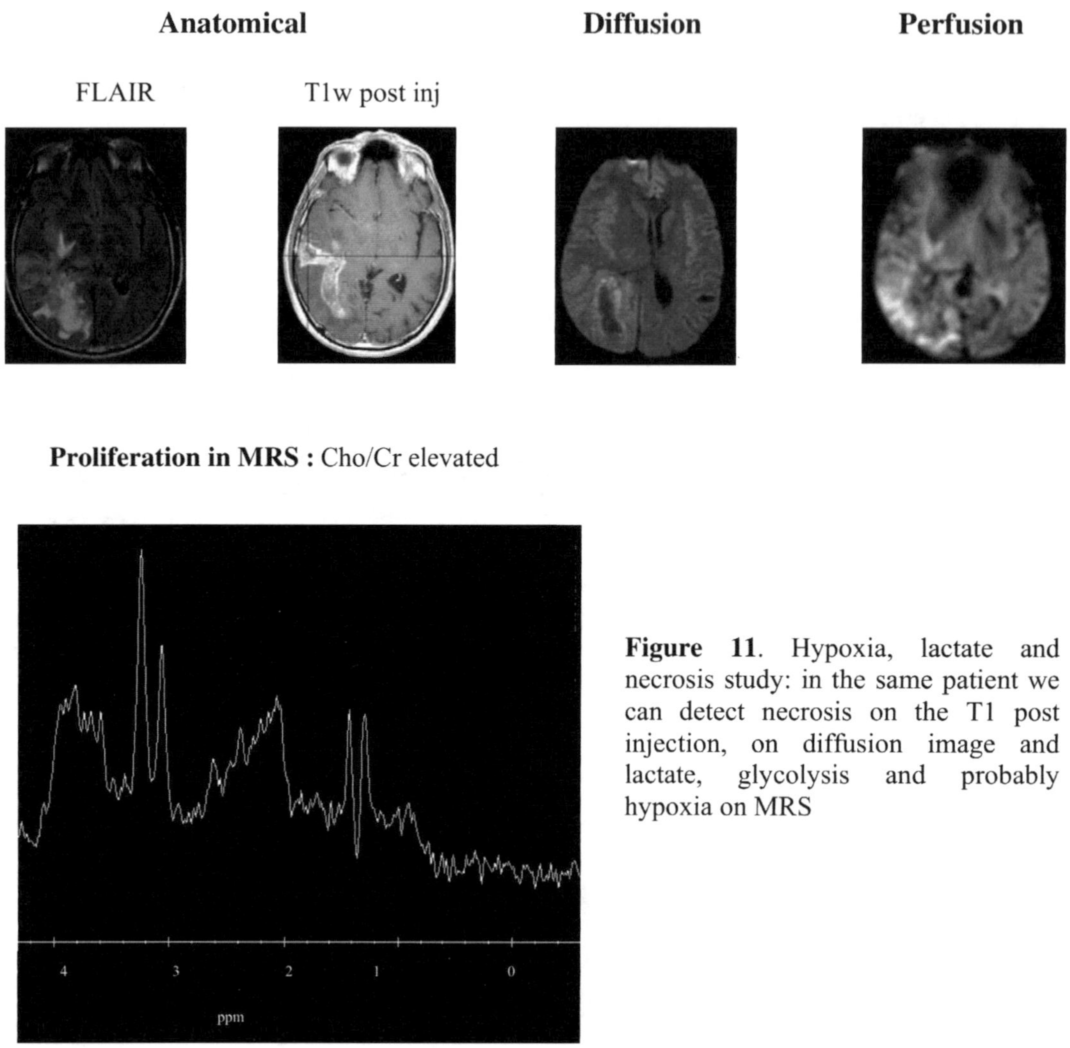

Figure 11. Hypoxia, lactate and necrosis study: in the same patient we can detect necrosis on the T1 post injection, on diffusion image and lactate, glycolysis and probably hypoxia on MRS

3. 5 Interest, objectives and example of the longitudinal MRS study

The longitudinal study allowed us to better understand therapeutic response in neurosciences or cancer, especially with regard to oxidative stress metabolic steps and therapies. The aim was to try to detect MRS differences due to in infiltration, proliferation, and relevant metabolite before any change in perfusion or contrast enhancement. The aim was also to see the tissue response to therapy. These studies could enable us to improve our understanding concerning the relationships between the different stages of tumoral processes with free radicals production and key metabolites measured by MRS like pyruvate, glutathione, citrate, succinate, acetate or hydroxyglutarate. In addition, those studies could improve the different prognostic classifications and the assessment of tumoral responses to different therapies and their optimal combinations.

Figure 12 is an example of a longitudinal study to demonstrate glial tumor metabolism and post chemotherapy variation and to determine cerebral variation in MRI and MRS area, amplitude, ratios of metabolites and spectral profiles during a 5 years prospective longitudinal follow-up of 24 patients (12 oligodendroglial tumours and 12 gliomatosis cerebri) treated with Temodal ® [31].The aggregation of different data on the box plot are not significant and not adequate. Therefore, the best way is to analyze the data longitudinally to be able to decrease variability and detect small changes earlier. The Cr concentration is quite stable in most of the oligodendrogliomas and in the majority of gliomatosis.

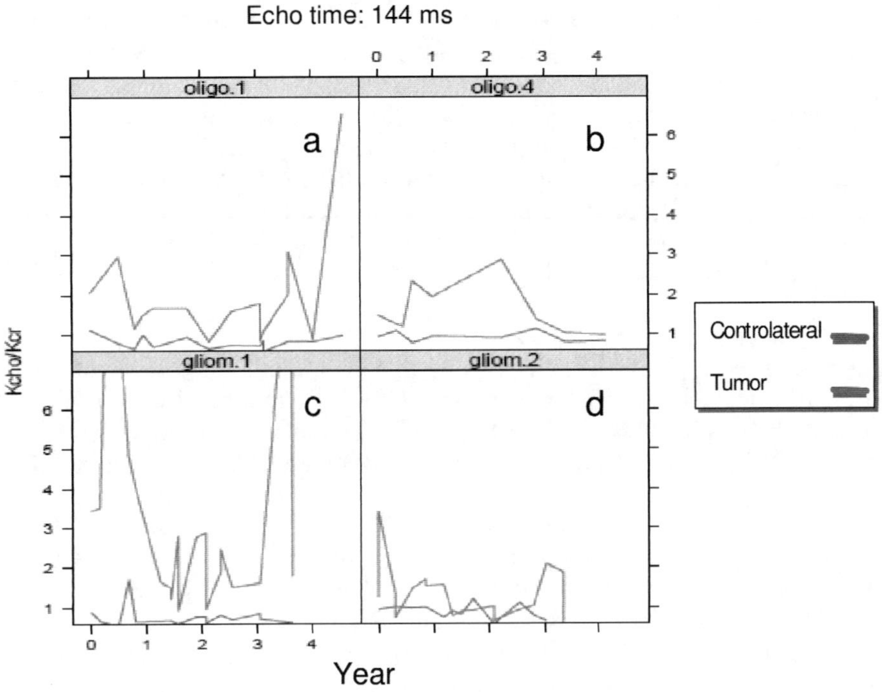

Figure 12 : Follow-up of Cho/Cr over at least 40 months during chemotherapy in spectra at echo time 144 ms: a) the first patient (a oligodendroglioma) is stable (Cho/Cr<2) until a progression with high proliferation (Cho/Cr>6) b) the second patient (also a oligodendroglioma) has aggravation (increase of Cho/Cr ratio until 3) and after treatment a decrease c) the third patient (a gliomatosis) have a very high and short period of proliferation (Cho/Cr>7) and then a decrease under treatment with some variability (Cho/Cr<3) and then again a high and short period of proliferation (Cho/Cr>7) d) the fourth patient has initial proliferation (Cho/Cr around 3) and then stability.

4. Discussion and future works

ROS-induced processes and metabolic switches in cancer could be measured by MRS. The aim is to enable us to study spontaneous ROS effects and other specific effects under therapy.

4.1 Antioxidants

Exposure to oxidative species has led organisms to develop a series of defence mechanisms, eg, antioxidants playing a protective role against oxidative stress by scavenging ROS. Enzymatic antioxidants include superoxide dismutase (SOD), glutathione peroxidase (GPx), and catalase (CAT) [32]. Non -enzymatic antioxidants include ascorbic acid (Vitamin C), alpha-tocopherol (Vitamin E) or glutathione (GSH) [34]. Under physiologic conditions, there is a balance between both the activities and the intracellular levels of these antioxidants. The introduction of antioxidants can inhibit tumor cell proliferation by modulating Receptor Tyrosine Kinase (RTK)-related intracellular signaling pathways [33] while ROS can stimulate neoangiogenesis, phospholipase c, protein kinase (PKC), gene regulation, and cell proliferation [20].

Anticancer drugs have been shown to induce apoptosis. Deficiencies in the apoptotic pathways may lead to chemoresistance. Studies report that glutathione (GSH) plays a critical role in the activation of apoptosis pathways due to anticancer drugs and conclude that dominant apoptosis resistance could depend on intracellular GSH levels with low GSH leading to apoptosis [35]. The GSH amount measured in vivo by MRS may help estimate antioxidant capacity and, indirectly, oxidative stress in tumors. Pyruvate is used as a precursor for glutathione synthesis and favors the restoration of the glutathione pool. These antioxidant properties of pyruvate and GSH (figure 8), while improving the cellular response and resistance to hypoxia might, however, interfere with anti-tumor treatment. In such cases, their measurement *in vitro* and *in vivo* using MRS might be an advantageous way to try to understand the processes induced by ROS in tumor cells and perhaps increase the effectiveness of some cancer therapies by predicting the resistance to anti-tumor treatment [36]. In addition, the measurement of glutathione levels in vivo may also be of potential relevance for measurements related to neurological disease processes and also viral infections.

4.2 Krebs cycle cetabolites involved in ROS regulation

Measurement of metabolites and processes by MRS could allow the study of oxidative stress and the redox balance. In fact, oxidized glutathione (GSSG) is reduced back to GSH by the enzyme glutathione reductase, which uses NADPH as the electron donor, cytosolic enzymes such as isocitrate dehydrogenase (IDH1) [37], and Glucose-6-Phosphate dehydrogenase [38] (figure 1).
Mitochondrial enzymes with modified activity (by mutations as with alpha-ketoglutarate dehydrogenase [39]) are coupled with NAD+/NADH, which can regulate ROS formation in the respiratory chain. We could then suggest that using MRS to identify and measure the quantity of metabolites involved in the formation of ROS and GSH (especially those from the Krebs cycle seen in blue in figure1) could reflect the balance between ROS and antioxidant formation involved in some diseases and tumor processes. This would contribute to our understanding of the role of ROS, in particularly in cancer. Therefore, MRS makes possible in vivo, non invasive and reproducible quantification of these metabolites indirectly involved in the regulation of ROS and the quantification of antioxidant (GSH) production. In addition, it is known that succinate oxidation can support the highest rate of ROS production and that accumulation of succinate due to gene mutation (e.g. succinate dehydrogenase (SDH)) [13] might also contribute to an increase in mitochondrial ROS generation. Therefore the quantification of succinate in patients could also be correlated with the ROS production, especially in abscesses.

4.3 Mitochondrial tumor tuppressor genes

The gene roles in tumorigenesis are critical. Their mutations are involved in the stabilization of HIF-1, in the up-regulation of genes involved in angiogenesis, in glucose transport, glycolysis or inhibition of developmental apoptosis. Recently, mutations of some enzymes such as isocitrate dehydrogenase (IDH1-IDH2), succinate dehydrogenase and fumarate hydratase were observed which could lead to accumulation of some substrates (succinate and fumarate) [13] to regulate NADH followed by changes in the production of ROS and antioxidant.

4.4 Relationship between ROS, HIF and hypoxia

It has been suggested that ROS, via changes in the cellular redox state, may trigger signalling pathways resulting in the activation of HIF-1 transcription factor. Some research is supported by the fact that tumors often have hypoxic areas that activate HIF1. This HIF-1 activation plays a key role in cancer development including metabolic switch by glycolytic enzyme activation, angiogenesis, or proliferation. However some studies show that tumor cells activate HIF-1 despite the absence of hypoxia. Involvement of ROS may explain this. Some studies have shown that mitochondria could regulate the stability of HIF through the increased production of ROS [40,41]. In addition, following radiotherapy, tumor reoxygenation leads to nuclear accumulation of HIF-1 in response to ROS, regulated by cytokines that could enhance cell radioresistance [42].

Journal of Physics: Conference Series **261** (2011) 012011 doi:10.1088/1742-6596/261/1/012011

4.5 HIF involvement in tumoral metabolism

Induction of HIF-1 leads to adaptive mechanisms to reduce ROS, especially uncontrolled mitochondrial generation of ROS. Under hypoxic conditions, perturbation in electrons flows from the respiratory chain results in increased ROS production due to ineffective electron transfer in the mitochondria. It would be toxic to cells if the TCA cycle flow would not be attenuated. The hypoxia inducible pyruvate dehydrogenase kinase 1 (PDK1) [43] is critical for the attenuation of mitochondrial ROS production and maintenance of ATP levels. The gene encoding for PDK1 is a direct target of HIF-1. PDK1 inactivates the pyruvate dehydrogenase (PDH) enzyme complex that converts pyruvate to acetyl-coenzyme A, thereby inhibiting pyruvate metabolism via the TCA cycle. HIF-1 plays three critical roles in the hypoxia-induced metabolic switch from oxidative to glycolytic metabolism [43,44,45]: (1) HIF-1 induces expression of upstream glucose transporters and glycolytic enzymes to increase flow from glucose to pyruvate, (2) HIF, via PDK1, blocks the conversion to acetyl CoA, and, (3) stimulates lactate dehydrogenase A to convert pyruvate to lactate. The induction of PDK1 seems necessary to prevent excessive mitochondrial ROS production and to shunt pyruvate transformation into lactate and regenerate NAD+, which allows glycolysis and ATP production under hypoxia. Concentrations of lactate, pyruvate, succinate, glucose and other metabolites involved in the TCA cycle can be estimated by MRS and allow better understanding of these processes in vivo with all the pathways regulations in a repeatable non destructive way.

Lactate (a doublet at 1.33 ppm) is observed in tumors resulting from aerobic and anaerobic glycolysis. Correlation was not always found between lactate levels and the grade of the tumor or metabolic rate because lactate occurs both in intra- and extra-cellular spaces and its overall level is a function of metabolic rate and clearance. In addition, lactate may pool in necrotic or cystic regions.

The presence of fatty acid synthesis (e.g., in necrosis and some membrane turnover) implied the need for two supporting pathways: (1) a source of NADPH (the electron donor for fatty acid synthesis), and, (2) an anaplerotic mechanism to replenish the TCA cycle intermediates during citrate export in order to meet energy needs during tumoral growth [46]. Glutamine metabolism can potentially fulfill both these needs in proliferating glioblastomas. In addition, HyperGln is easily detectable by MRS in high-grade gliomas, increased cell cytolysis, mitochondrial disease, hyperammonemia, and osmotic and hepatic encephalopathy.

4.6 ROS and angiogenesis

Studies found formation of excessive amounts of ROS after radiation treatment (actually it is the aim of this treatment) and activation of HIF-1 during the reoxygenation, which leads to upregulation of VEGF and other HIF-1 regulated pro-angiogenic factors.

Many studies have shown that ROS production occurs as a result of activated growth factor receptor signalling such as receptor-tyrosine kinase. ROS generated by these pathways can function as true second messengers and mediate important cellular functions such as proliferation and programmed cell death [47]. Following radiotherapy, tumor reoxygenation leads to nuclear accumulation of HIF-1 in response to ROS, which might generate enough oxidative stress to stabilize HIF-1. The resulting increase in HIF-1 regulated cytokines enhances cell radioresistance because the HIF-1 complex regulates the expression of more than 40 genes involved in tumor metabolism, growth, and angiogenesis, including VEGF [48]. The effect of ROS on angiogenesis could be illustrated by MR perfusion data.

4.7 Proliferation

HIF-1 activation via ROS production may promote tumor proliferation by angiogenesis and metabolic switches. The oxidative stress also plays a role in the activation of PKC isoforms and cell proliferation [20]. The tCho peak is seen *in vivo* due to rapid membrane synthesis of dividing cancer cells and is predominantly due to PC (a precursor for cell membrane synthesis) and GPC (involved in cell membrane degradation). tCho levels were shown to correlate with proliferative potential in gliomas.

In addition to its main roles, methionine could have an anti-oxidant role by protecting cells from free radical damage [49]. Methionine could protect against lipid peroxidation, membrane damage, and changes in the glutathione system. On the other hand, 11C methionine PET tracer (as seen p. 9 in *PET and Multimodality*) [50,51] could be used to study tumor proliferation in relation with the Cho/Cr MRS parameter.

4.8 ROS – Necrosis - Apoptosis

Lipids (1.3 and 0.9 ppm) are seen in high-grade tumors [11]. Kuesel et al., from ex vivo studies, showed that lipids correlate with necrosis, a histological characteristic of high-grade tumors [12]. Increased lactate and lipids are seen with glioma grades that represent the progression from viable tumor cells to severe hypoxia, followed by necrosis. Recent studies showed that MRS spectral lipid profiles contain diagnostic information and may be used to differentiate glioblastomas from metastases [10]. Absolute lipid concentration could also help to differentiate toxoplasmosis from lymphoma at certain stages. ROS can be considered as mediators of cell death (by apoptosis or necrosis) and ROS can activate anti-apoptotic pathways leading to increased cellular survival. For example Bcl-2 family members are involved in control of ROS or ROS-related processes [52]. MRS could improve our understanding of the physiopathologic mechanisms of necrosis and apoptosis, determine whether their origin is infectious or tumoral, or provide information about the rate of development of these processes.

4.9 Infiltration

Reduction of NAA (found in neurons) in adult brain tumors is non-specific, as it also occurs in other brain pathologies with neuronal damage such as strokes or MS. The presence of NAA in a tumor may indicate viable neurons within an infiltrative tumor such as a glioma. The parameter, choline-NAA-index (CNI), very robust and related to high grade, is used with some success for defining the regional abnormalities in multivoxel spectroscopy of gliomas [4]. NAA/Cr is also an important ratio to follow in the contralateral side for the assessment of therapeutic effectiveness.

4.10 Under therapy

Radiation and chemotherapy and antiangiogenic therapies are widely used in the treatment of these malignant brain tumors. The limitation of brain radiation therapy and chemotherapy in humans not only comes from the inability to kill certain types of tumor cells and to induce cellular death in different types of environmental conditions (eg, pH, hypoxia, perfusion, energy, necrosis, cell density, and edema) but also from excessive damage to normal appearing brain which increases cognitive impairment. Among the types of radiation-induced brain toxicities, late delayed effects together with the original perturbations [53] can lead to severe and irreversible neurological dysfunctions (assessed by NAA/Cr). Following radiation, late delayed effects of exposure within the central nervous system have been attributable to both parenchymal and vascular damage involving oligodendrocytes, glial cells, neural progenitors, neurons and endothelial cells. This dynamic process, involving radiation-induced damage of target cells, could sometimes involve secondary reactive neuroinflammatory and glial processes that could lead to cell loss, tissue damage, metabolic changes, and functional deficits. The progressive, delayed processes and damage to the brain and the tumors after high-dose radiation and chemotherapy can be caused by radiation-induced long-lived processes induced by free radicals, reactive oxygen species, and pro-inflammatory cytokines. To better understand and follow longitudinally the metabolic effect of ROS after clinical radiotherapy (conventional, gamma-knife, protontherapy and heavy ions) is important.

New drug types, their design and combination therapies, especially those that could exert their effects by blocking pro-inflammatory cytokines and reactive oxygen species (eg, glutathione) and avoid irreversible metabolic switches, could be studied in future clinical trials.

5. Conclusion

(MRI) and spectroscopy (MRS) and perfusion could be useful for monitoring the Reactive Oxygen Species (ROS) effects on energy and glycolytic metabolism, mitochondrial function, lipid metabolism, necrosis, apoptosis, cell proliferation, and infiltration. These changes, as shown, could be studied longitudinally and non-invasively in humans with glial brain tumors, abscesses, or mitochondrial disease after therapy. Despite the difficulty, the variability, and unknown factors, these repeated measures could give us more insights into the evolution at different times of the pathological processes, stages, tumor progression and response to therapy.

Acknowledgement

The authors thank the MELUSYN network and Stéphane Allouche for fruitfull discussions. We are indebted to Louisa Barré, Stéphane Guillouet and Alain Manrique for providing the PET tracers and images.

References

[1] Galanaud D, Chinot O, Nicoli F, Confort-Gouny S, Le Fur Y, Barrie-Attarian M, Ranjeva JP, Fuentès S, Viout P, Figarella-Branger D and Cozzone PJ 2003 *J Neurosurg* **98** 269.
[2] Gill SS, Thomas DGT, Van Bruggen N, Gadian DG, Peden CJ, Bell JD, Cox IJ, Menon DK, Iles RA, Bryant DJ and Couttsd GA 1990 *J Comput Assist Tomogr* **14** 497.
[3] Seymour ZA, Panigrahy A, Finlay JL, Neson MD and Blumi S 2008 *Am. J.Neuroradiol.* **29**1006.
[4] Nelson SJ, Graves E, Pirzkall A, Li XJ, Chan AA, Vigneron DB and McKnight TR 2002 *J Magn Reson Imaging* **16** 464.
[5] Ross B, Michaelis T 1994 *Magn Reson Q* **10** 191.
[6] Opstad KS, Provencher SW, Bell BA, Griffiths JR and Howe FA 2003 *Magn Reson Med* **49** 632.
[7] Cho YD, Choi GH, Lee SP and Kim JK 2003 *Magn Reson Imaging* **21** 663.
[8] Rijpkema M, Schuuring J, van der Meulen Y, van der Graaf M, Bernsen H, Boerman R, van der Kogel A and Heerschap A 2003 *NMR Biomed* **16** 12.
[9] Stubbs M, Veech RL and Griffiths JR 1995 *Adv Enzyme Regul* **35** 101.
[10] Howe FA and Opstad KS 2003 *NMR Biomed* **16** 123.
[11] Auer DP, Gössl C, Schirmer T and Czisch M 2001 *Magn Reson Med* **46** 615.
[12] Kuesel AC, Sutherland GR, Halliday W and Smith IC 1994 *NMR Biomed* **7** 149.
[13] King A, Selak M A and Gottlieb E 2006 *Oncogene* **25** 4675.
[14] Thompson C 2009 *New Engl. J. Med.* **360** 813.
[15] Dang L, White D, Gross S, Bennett B, Bittinger M, Driggers E, Fantin V, Hyun Gyung Jang, Shengfang J, Keeenan M, Marks K., Prins R, Ward P, Yen K, Liau L, Rabinowitz J, Cantley L, Thompson C, Vander Heiden M and Su S 2009 *Nature* **465** 966.
[16] Heinzelmann S and Bauer G 2010 *Biol. Chem.* **391**, 675.
[17] Poyton R, Ball K and Castello P 2009 *Cell Press, TrendsEndocrinol Metabol.* **20** 7.
[18] Valko M, Leibfritz D, Moncol J, Cronin MTD, Mazur M and Telser J 2007 *Inter. J. Biochem. Cell Biol.* **39** 44.
[10] Dreher D and Junod AF 1996 *Eur. J. Cancer* **32A** 30.
[20] Storz P 2005 *Frontiers in Bioscience* **10** 1881.
[21] Dou W, Ruan S, Chen YP, Blovet D and Constans JM 2007 *Image Vision Comput* **25** 164.
[22] Dou W PhD Thesis, Caen University 2006.
[23] Constans JM Thesis, Caen University, 2006.
[24] Delcroix N PhD Thesis, Caen University 2000.
[25] Chechin D PhD Thesis, Caen Univesrity 2001.
[26] Provencher S 1982 *Comput. Phys. Commun.* **27** 213.
[27] Nelson S 1989 *J. Magn. Res. Imag.* **84** 95.

[28] Naressi A 2001*Computers Biol. Med.* **31** 269.
[29] Hossu G PhD Thesis, Caen University 2009.
[30] Guo Y, Ruan S, Landre J and Constans JM 2010 *IEEE Trans Biomed Eng.* **57** 1620.
[31] Constans JM, Hossu G, Kauffmann F Dou W, Ruan S, Rioult F, Derlon J M, Lechapt Zalcman E, Bernaudin M, Chapon F, Valable S, Courthéoux P and Guillamo J S 2009 *17th ISMRM* Honolulu USA.
[32] Bechtel W, Bauer G 2009 *Anticancer Res.* **29** 4541.
[33] Martin V, Herrera F, Garcia-Santos G, Antolin I, Rodriguez-Blanco J and Rodriguez C 2007 *Free Rad.l Biol. Med.* **42** 1715.
[34] Choi I and Gruetter R 2004 *J. Neurochem.* **91** 778.
[35] Friesen C, Kiess Y and Debatin KM 2004 *Cell Death Differ.* 11 S73.
[36] Roudier E and Perrin A 2009 *Biochim. Biophys. Acta* **1796** 55.
[37] Lee SM, Koh HJ, Park DC, Song BJ, Huh TL and Park JW 2002 *Free Rad. Biol .Med.* **32** 1185.
[38] Salvemini F, Franze A, Iervolino A, Filosa S, Salzano S and Ursini MV 1999 *J.Biol. Chem.* **274** 2750.
[39] Tretter L and Adam-Vizi V 2004 *J. Neurosci.* **24** 7771.
[40] Lopez-Lazaro M 2009 *Anti-Cancer Agents Med. Chem.* **9** 517.
[41] William G and Kaelin Jr 2005 *Cell Metabolism* **1** 357.
[42] Moeller BJ, Cao Y, Li CY and Dewhirst MW 2004 *Cancer Cell* **5** 429.
[43] Tabatabaie T, Potts J and Floyd R 1996 *Arch. Biochem. Biophys.* **336** 290.
[44] Semenza G 2010 *Curr.Opinion Gen.Dev.* **20** 51.
[45] Simon M C 2006 *Cell Metabolism* **3** 150.
[46] DeBerardinis RJ, Mancuso A , Daikhin E, Nissim I, Yudkoff M, Werhli S and Thompson CB 2007 *Proc Natl Acad Sci U S A* **104** 19345.
[47] Ushio-Fukai M and Nakamura Y *2008 Cancer Lett.* **266** 37.
[48] Dewhirst M W, Cao Y T, Li C Y and Moeller B 2007 *Radiotherapy and Oncology* **83** 249.
[49] Bobrowski K, Houée Levin C and Marciniak B 2008 *Chimia* **63** 728.
[50] Derlon JM, Bourdet C, Bustany P, Chatel M; Theron J, Darcel F and Syrota A 1989 *Neurosurgery* **25** 720.
[51] Constans JM, Soussan L, Fernandez J, Chapon F, Derlon JM and Courthéoux P 1997 *Med. Biol. Eng. Comp.* **35** 764.
[52] Burkitt M J and Wardman P 2001 *Biochem. Biophys. Res.Com.* **282** 329.
[53] Lacombe S, Sabatier L, Wien F and Gauduel YA 2010 *Cell Death Dis.* **1** e4.

Journal of Physics: Conference Series **261** (2011) 012012 doi:10.1088/1742-6596/261/1/012012

Ion microbeam irradiation for radiobiology and radical chemistry: status and prospect

H Khodja
CEA, IRAMIS, SIS2M, LEEL, 91191 Gif-sur-Yvette, France.
CNRS, UMR 3299, SIS2M, LEEL, 91191 Gif-sur-Yvette, France.

E-mail: hicham.khodja@cea.fr

Abstract. Ion microbeams are commonly used to study local irradiation effects in living cells, as it has been established that ion beam irradiations can lead to deleterious changes in cells that are not struck directly by the microbeam. Such changes, which take place over distances long compared to the size of the irradiation spot and for times long compared to the time of irradiation, are collectively termed radiation-induced bystander effect or RIBE. Free-radical chemistry is frequently invoked to explain the RIBE but no unified model is available at present. Ion microbeams when coupled with advanced methods for observing free radicals are the tools of choice for investigating the chemistry and biological processes governing RIBE.

1.Introduction

Intense efforts have been made over the two last decades to clarify the ways that living cells, tissues, and organisms respond to low dose radiations. Indeed, many experiments have shown that a low dose of ionizing radiation produces deleterious effects not only directly in the cells targeted, but also indirectly in cells that receive no direct dose. Such changes are known collectively as the radiation induced bystander effect (RIBE). The RIBE has practical implications for radiation protection and for the administration of radiotherapy. Today, the RIBE seems to be uncontroversial for large linear energy transfer (LET) irradiations when the dose is administered with light ions (protons, α-particles). These irradiations are representative of environmental radiation hazards, and in certain cases are used for radiotherapy purposes. The term 'bystander effect' is now also employed to describe secondary effects on neighboring cells induced by other kinds of stresses, namely, chemical [1], heat shock [2], and nanoparticle absorption [3]...). Today many laboratories around the world are engaged in investigating the chemical mechanisms and kinetics of the RIBE. Often, the experiments include a device for producing a microbeam of irradiation [4] for its ability of delivering a controlled dose in a spatially restricted area.

Several lines of experimental evidence demonstrate the essential role of free radicals in the appearance and propagation of RIBE. Although neither the underlying mechanisms nor the full range of observed changes are fully understood today, the study of free-radical chemistry has progressed during these last years [5], and has led to an improved description of radical formation and evolution in solutions following an irradiation event. Particularly, interactions between radical species are well understood and models with a high reproducibility have been developed. Both experimental methods and efficient simulation tools are available today and should contribute to the emergence of comprehensive models for the RIBE.

2. Radiation induced Bystander effect

Several comprehensive reviews regarding RIBE are available in the literature [6-8]. The existence of RIBE was first demonstrated nearly 20 years ago [9] when it was reported that after irradiating with 3.7 MeV α-particles only 1% of cell nuclei, more than 30 % of the cell population exhibited sister chromatid exchanges, a mutagenic event. This work showed that a direct interaction between energetic particles and nucleus is not needed to induce deleterious effects in cells. Later it was established that even in the case of cytoplasmic irradiations, targeted cells were damaged and RIBE was observed in neighboring cells [10,11]. Recently, it was demonstrated that pre-cancerous cells may be selectively directed to apoptosis (genetically programmed suicide) through RIBE by irradiating neighboring normal cells at relatively low doses [12]. Other experiments demonstrated that cells located up to 1 mm from the initial impact of ionizing radiation exhibit DNA damages [13,14]. Today the RIBE has been demonstrated in a large panel of cell lines and has been observe with different types of radiations, including x-rays, γ-rays and heavy ions, but recent reports have questioned its effectiveness when irradiating with x-rays and heavy ions [15,16]. It seems however uncontroversial that RIBE is clearly induced when using α particles [15].

Quantitative and semi-quantitative methods are applied to measure the induction of deleterious effects on cells and identify associated pathways, and are generally based on fluorometric measurements: fluorescence of DNA repair protein binding antibodies, observation of cellular radical induction and amplification by direct or indirect (scavenging) techniques. Modulation of gene expression is also employed to identify the biological mechanisms governing the RIBE: deficient cell lines in terms of DNA repair or inter-cellular junction potency have been exposed to microbeam irradiations, suggesting an intercellular communication process through gap junctions [17]. A typical RIBE experiment is presented in figure 1.

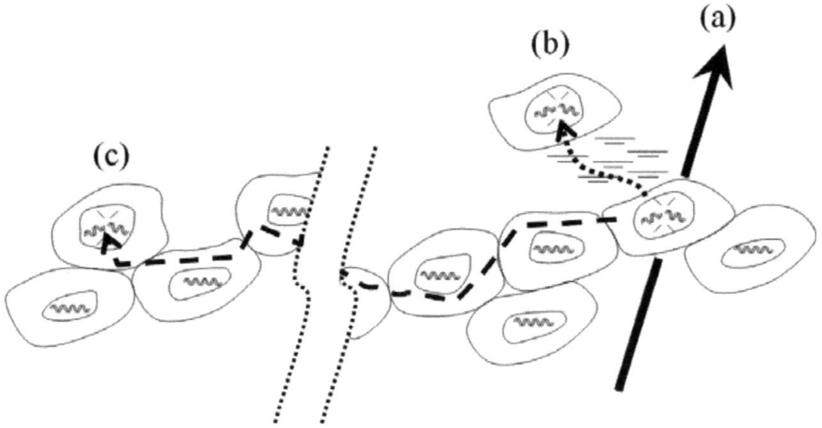

Figure 1. Typical RIBE in cultured cells. (a): incoming ionizing radiation, (b) RIBE in neighboring cells mediated through culture medium, (c) RIBE in distant cells mediated through gap junctions.

Very early, radical species were suspected to trigger and also mediate the observed RIBE. The introduction of radical scavengers such as DMSO inhibits the RIBE. Scavengers however interfere with normal signaling activity among cells. In some experiments [18,19], "direct" observation of radical species, based on fluorescent radical probes confirmed the link between radical and RIBE spatial distributions. However fully quantitative evaluations using radical fluorescent probes are controversial as it has been shown that probe oxidation induces intermediate radical creation, leading to emergence of a self-amplified signal [20].

3. Microbeams for RIBE investigation

Interactions between the ion beam projectiles (e.g. protons) and atomic nuclei in a target produce recoil ions and nuclear fragments and ionize bounded electrons. Spectroscopy of all these processes is used to obtain information about element concentrations. Initially, ion microbeams (mainly protons and less often deuterons, alpha particles, and heavier ions accelerated to energies of ~ 1 MeV/nucleon) were employed for elemental analysis of solid samples with dimensions of tens of micrometers [21]. As the methods improved, elemental mapping with micrometer-scale resolution became possible.

Ion microbeams are a good tool to induce the RIBE and to control its magnitude. The beams can be chosen and tuned to generate ion tracks in tiny regions (typically less than few microns) near cell surfaces. The pioneering work was done from 1953-1963 (see [22]). During the 1990s, groups at Gray Cancer Institute [23] and Radiological Research Accelerator Facility [24] built facilities which demonstrated the interest of microbeam setups for RIBE investigations. Since then, about twenty laboratories started investigating this new field. Recent reviews of the facilities used for cellular irradiation are given in [25,26]. Most workers use experimental setups that ideally would include the following elements (see figure 2): (i) a sub-micrometer ion beam; (ii) a horizontal sample stage that can be operated under atmospheric pressure; (iii) targets consisting of a cell monolayer cultured with fresh medium and located in a standard culture environment, (iv) a device for precisely controlling the number of charged particles delivered to the target; (v) a microscope for observing the targets optically; (vi) tools for observing changes in the target.

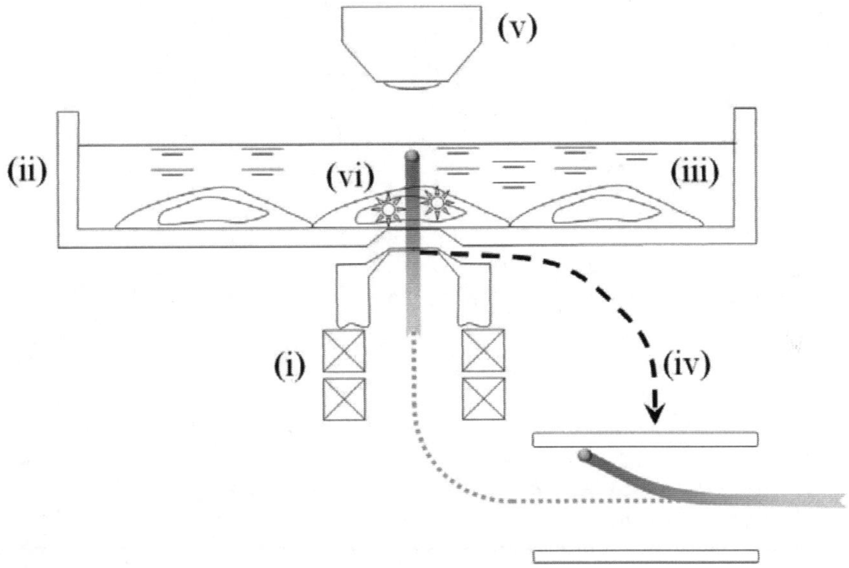

Figure 2. An ideal experimental setup for RIBE investigation. For explanations of parts (i)-(vi), see text.

System elements (iv-vi) must have minimal impact on biochemical processes. The listed requirements are however difficult to reach concomitantly and are frequently conflicting. Sub-micrometer beams are available only under vacuum. If the target is to be irradiated in air, passage of the beam through a window introduces unavoidable radial and angular scattering and inevitably enlarges the impact area. The smallest reported beam spots on target are about 1 μm [27]. Many ion beam facilities were originally devoted to ion beam analysis with beamlines mounted horizontally, requiring targets to be mounted vertically. A growing number of facilities have built the complex hardware needed to bend the beam in order to irradiate horizontal targets [28,29]. Different strategies are used to achieve the unusual culture conditions, either by the use of moisturized atmosphere or micro-environmental chamber [30,31]. Ion counting can be performed by detecting ion crossing through a scintillator thin film located between the exit window of the accelerator and the target cells, but at the cost of increasing beam scattering and consequently reducing the hit precision of the beam on target. Alternatively, ions can be counted by using a charged particle detector positioned 'downstream', i.e., behind the targeted cells. This approach requires the removal of the culture medium, which otherwise would prevent the beam particles from reaching the detector. Removal of the culture medium may induce potential parasitic stresses on the cells.

Cell observation is intrinsically a disturbing process, and a simple operation such as tagging a nucleus with a fluorescent probe in order to make it observable, can lead to misinterpretation if used without great care [32]. As already pointed out, fluorometric behavior of radical probes has been intensively discussed. Some probes such as the popular 2',7'-dichlorofluorescein are suspected to induce radical generation leading to signal amplification or are known to be photosensitive [20,33].

4. Radical chemistry: local levels

The aqueous chemistry of radical species produced in cells by ionizing radiation is important in many fields, particularly in radiobiology. These radical species evolve with complex spatial and temporal dependencies, owing to the large range of reaction rates (10^{-3} to 10^{1} M^{-1} s^{-1}) and diffusion coefficients (10^{-5} to 10^{-6} cm^2 s^{-1}) [34]. Radical chemistry has been intensively investigated for decades, and many experimental and simulation tools have been developed to elucidate the impact of radicals on molecules of interest in biology (DNA, proteins). The systems studied include both dry extracts or solutions. Although the "direct/indirect effect" terminology employed in this literature could evoke RIBE, such studies are in fact conducted over very small distance scales, typically less than 1 μm, i.e., for the immediate molecular environment of the interacting species. The temporal behavior of generated species has been investigated by the combined techniques of pulse radiolysis and radical scavenging. A clear dependency of main generated radical species type with radiation quality is evidenced [35]. Spatial recombination of the different generated free radicals species, as far as we know, has not yet been observed experimentally. Numerous simulation reports treat the interactions of free radicals, but the maximum interaction distance considered to date is less than 1 μm.

In the radical chemistry approach, DNA and protein damages are treated as a consequence of interactions between the molecules in the sample and the incoming particle during its deceleration. The swift particles (primary ions or secondary electrons) can affect chemical bonds in the target by ejecting bonding electrons or by exciting them to states in which molecules may rearrange in undesirable ways. Typically, damages as double strand breaks or even more complex damages such tandem damages are considered as main deleterious cell nucleus damages as repair systems are not always fully efficient. Obviously, these effects are inherently spatially limited to the track volume. This approach is far too limited in spatial extent to explain the RIBE effect. The energy of secondary electrons emitted along an ion track is governed by the maximum energy transfer rule:

$$E_e^{Max} = 4 \times E_i \frac{m_e m_i}{(m_e + m_i)^2} \tag{1}$$

where E_e^{Max} is the maximum transferred energy to electrons, E_i is the ion kinetic energy, m_e and m_i being electron and ion masses, respectively. Typically, this leads to transferred electron kinetic energies lower than 2 keV. Electrons with energies in this range cannot reach neighboring cells or even the cell nucleus in case of cytoplasmic irradiation. Pimblott et al [36] calculated that for various ion projectiles (H^+, α, C^{6+}), the most probable secondary electron energy is 10 eV, with the mean at 60 eV, leading to a projected range of approximately 1 nm in water.

An exhaustive study regarding spatial and temporal distributions of radical species induced by a scanning electron beam with water was reported by Royall et al [37]; they conclude that the remaining radical species are hydrogen peroxide and hydroxyl for dwell times up to 1 ms. For longer dwell times, a quasi-steady state is reached for which spatial repartition is governed by recombination and diffusion processes. In this regime, radical species are found much farther from the track vicinity (up to 15 μm from central track with 25 keV electron beam). The main contributors to radical production are electrons from the primary beam that have slowed to thermal velocities, but continue to diffuse through water. It is interesting to note that the relatively large spatial extent of the radical distribution is not associated experimentally with RIBE induction with electron beams [38]. In a tentative Monte-Carlo investigation of RIBE, Muroya et al [39] investigated time and spatial distributions of species generated following ion irradiation with 7-MeV α-particles. The dominant species, hydrogen peroxide, is found uniformly distributed in a 1 μm-diameter cylindrical tube 10 μs following the arrival of the ion. It is however important to note that the study was limited here to a 1 μm-section of the track. Plante et al [40] looked recently at radical and dose radial distributions for heavy ion tracks (10 to 170 keV/μm) by means of Monte-Carlo simulations; no detectable events were found outside a 100 nm cylinder.

The most probable contribution to the production of radical species outside the immediate zone of interaction has to be associated with the lateral straggling that affects the primary ion path in the Bragg region where nuclear collisions are dominant. This leads to an uncertainty of the track path localization. Typically, in the case of a 3 MeV α-particle beam, lateral ion straggling in water estimated using SRIM software [41] is 430 nm, while no significant additional contribution is associated with secondary electrons emitted in this part of the track owing to the low energy transfer occurring at the end of the path. These data confirm that radical generation through radiolysis process cannot directly induce signaling on neighboring cells. Consequently, deleterious effects observed in non targeted cells by the RIBE process are necessarily initiated by biochemical radical species and are highly dose distribution dependent.

5. Future directions

Today huge amounts of data regarding RIBE in numerous cell lines have been accumulated by means of microbeam techniques. Although radical species are clearly identified as major actors in RIBE, physical chemical considerations exclude bystander induction by purely chemical processes, implying mandatory biological steps. Experimental protocols vary, however, and it has not yet been possible to create a unified, quantitative model of RIBE. Joint efforts from radiobiology and radiation chemistry communities are needed. Recent progress in detecting small amounts of radical species in small regions should open new perspectives. In particular, fluorescent nanomaterials, such as nanoparticles [42] and nanotubes [43] are promising tools, as signals from a single molecule now can be detected. Encoded fluorescent proteins can be also used to map the distribution of radical species in cells [44].

Improvements in beam targeting precision, with minimal parasitic ion detection should be helpful in controlling the dose microdistribution in cells. In situ observation methods for cell nucleus identification should be effective in a near future provided that they can be engineered not to have side effects. In parallel, microdosimetry simulations, (see [45]) might give hints about the effective cellular target where transition between chemical and biological stages occurs. As indicated above, RIBE is not systematically observed with different particles having the same LET. Thus, 3D modeling of the interactions between incoming ion and target material, particularly the end of the track during the last stages of ion deceleration, should be of major importance in understanding the underlying processes.

Acknowledgements

Author is grateful to G Herzog (Rutgers University NJ USA) for suggestions and helpful comments.

References

[1] Cogan N, Baird D M, Phillips R, Crompton L A, Caldwell M A, Rubio M A, Newson R, Lyng F and Case C P 2010 *Mutat. Res.-Fundam. Mol. Mech. Mutag.* **683** 1.

[2] Purschke M, Laubach H-J, Anderson R R and Manstein D 2009 *J Invest Dermatol* **130** 86.

[3] Bhabra G, Sood A, Fisher B, Cartwright L, Saunders M, Evans W H, Surprenant A, Lopez-Castejon G, Mann S, Davis S A, Hails L A, Ingham E, Verkade P, Lane J, Heesom K, Newson R and Case C P 2009 *Nat Nano* **4** 876.

[4] Schettino G, Al Rashid S T and Prise K M 2010 *Mutat. Res. - Rev. Mut. Res.* **704** 68.

[5] Aitken C E, Marshall R A and Pulglisi J D 2008 *Biophys. J.* **94** 1826-35

[6] Matsumoto H, Hamada N, Takahashi A, Kobayashi Y and Ohnishi T 2007 *J. Radiat. Res.* **48** 97.

[7] Prise K M and O'Sullivan J M 2009 *Nat Rev Cancer* **9** 351.

[8] Averbeck D 2010 *Mutat. Res.-Fundam. Mol. Mech. Mutag.* **687** 7.

[9] Nagasawa H and Little J B 1992 *Cancer Res.* **52** 6394.

[10] Shao C, Folkard M, Michael B D and Prise K M 2004 *Proc. Natl. Acad. Sci. USA* **101** 13495.

[11] Tartier L, Gilchrist S, Burdak-Rothkamm S, Folkard M and Prise K M 2007 *Cancer Res.* **67** 5872.

[12] Bauer G 2007 *Int. J. Radiat Biol.* **83** 873.

[13] Belyakov O V, Mitchell S A, Parikh D, Randers-Pehrson G, Marino S A, Amundson S A, Geard C R and Brenner D J 2005 *Proc. Natl. Acad. Sci. USA* **102** 14203-8

[14] Kashino G, Kondoh T, Nariyama N, Umetani K, Ohigashi T, Shinohara K, Kurihara A, Fukumoto M, Tanaka H, Maruhashi A, Suzuki M, Kinashi Y, Liu Y, Masunaga S-i, Watanabe M and Ono K 2009 *Int. J. Radiat. Oncol. Biol. Phys.* **74** 229-36

[15] Sowa M B, Goetz W, Baulch J E, Pyles D N, Dziegielewski J, Yovino S, Snyder A R, de Toledo S M, Azzam E I and Morgan W F 2010 *Int. J. Radiat Biol.* **86** 102-13

[16] Fournier C, Barberet P, Pouthier T, Ritter S, Fischer B, Voss K O, Funayama T, Hamada N, Kobayashi Y and Taucher-Scholz G 2009 *Radiat. Res.* **171** 530.

[17] Azzam E I, de Toledo S M and Little J B 2001 *Proc. Natl. Acad. Sci. USA* **98** 473

[18] Lyng F M, Maguire P, Kilmurray N, Mothersill C, Shao C, Folkard M and Prise K M 2006 *Int. J. Radiat Biol.* **82** 393 .

[19] Hanot M, Hoarau J, Carrière M, Angulo J F and Khodja H 2009 *Int. J. Radiat. Oncol. Biol. Phys.* **75** 1247.

[20] Wardman P 2007 *Free Radical Biol. Med.* **43** 995.

[21] Pierce T B, Peck P F and Cuff D R A 1969 *Nucl Instrum Methods* **67** 1.

[22] Zirkle R E and Bloom W 1953 *Science* **117** 487.

[23] Folkard M, Vojnovic B, Prise K M, Bowey A G, Locke R J, Schettino G and Michael B D 1997 *Int. J. Radiat Biol.* **72** 375.

[24] Randers-Pehrson G, Geard C R, Johnson G, Elliston C D and Brenner D J 2001 *Radiat. Res.* **156** 210-4

[25] Bigelow A W, Brenner D J, Garty G and Randers-Pehrson G 2008 *IEEE Trans. Plasm. Sci.* **36** 1424.

[26] Gerardi S 2009 *J. Radiat. Res.* **50** A13.

[27] Randers-Pehrson G, Johnson G W, Marino S A, Xu Y, Dymnikov A D and Brenner D J 2009 *Nucl. Instrum. Methods Phys. Res., Sect. A* **609** 294.

[28] Daudin L, Carrière M, Gouget B, Hoarau J and Khodja H 2006 *Radiat. Prot. Dosim.* **122** 310.

[29] Kobayashi Y, Funayama T, Hamada N, Sakashita T, Konishi T, Imaseki H, Yasuda K, Hatashita M, Takagi K, Hatori S, Suzuki K, Yamauchi M, Yamashita S, Tomita M, Maeda M, Kobayashi K, Usami N and Wu L 2009 *J. Radiat. Res.* **50** A29.

[30] Hable V, Greubel C, Bergmaier A, Reichart P, Hauptner A, Krücken R, Strickfaden H, Dietzel S, Cremer T, Drexler G A, Friedl A A and Dollinger G 2009 *Nucl. Instrum. Methods Phys. Res., Sect. B* **267** 2090.

[31] Khodja H, Hanot M, Carrière M, Hoarau J and Angulo J F 2009 *Nucl. Instrum. Methods Phys. Res., Sect. B* **267** 1999.

[32] Gault N, Rigaud O, Poncy J-L and Lefaix J-L 2007 *Radiat. Res.* **167** 551.

[33] McArdle F, Pattwell D M, Vasilaki A, McArdle A and Jackson M J 2005 *Free Radical Biol. Med.* **39** 651.
[34] Hill M A and Smith F A 1994 *Radiat. Phys. Chem.* **43** 265.
[35] Baldacchino G 2008 *Radiat. Phys. Chem.* **77** 1218.
[36] Pimblott S M and LaVerne J A 2007 *Radiat. Phys. Chem.* **76** 1244.
[37] Royall C P, Thiel B L and Donald A M 2001 *J. Microsc.* **204** 185.
[38] Morgan W F, Goetz W and Sowa M B 2009 *J. Radiat. Res.* **50** A81.
[39] Muroya Y, Plante I, Azzam E I, Meesungnoen J, Katsumura Y and Jay-Gerin J-P 2006 *Radiat. Res.* **165** 485.
[40] Plante I and Cucinotta F A 2010 *Radiat. Environ. Biophys.* **49** 5.
[41] Ziegler J F, Biersack J P and Littmark U 1985 *The stopping and range of ions in solids* (New York: Pergamon)
[42] Casanova D, Bouzigues C, Nguyen T-L, Ramodiharilafy R O, Bouzhir-Sima L, Gacoin T, Boilot J-P, Tharaux P-L and Alexandrou A 2009 *Nat Nano* **4** 581.
[43] Jin H, Heller D A, Kalbacova M, Kim J-H, Zhang J, Boghossian A A, Maheshri N and Strano M S 2010 *Nat Nano* **5** 302.
[44] Belousov V V, Fradkov A F, Lukyanov K A, Staroverov D B, Shakhbazov K S, Terskikh A V and Lukyanov S 2006 *Nat Methods* **3** 281.
[45] Incerti S, Seznec H, Simon M, Barberet P, Habchi C and Moretto P 2009 *Radiat. Prot. Dosim.* **133** 2.

In situ UV-visible spectrum acquisition of Br_3^-. Investigations of concentrated HBr aqueous solutions under 13-keV X-rays

D Saffré[1], E Atinault[1,2], S Pin[1], JP Renault[1], JL Hazemann[3] and G Baldacchino[1]
[1]CEA Saclay, IRAMIS, UMR3299 CEA/CNRS SIS2M, Laboratoire de Radiolyse, 91191 Gif-sur-Yvette Cedex, France - [2]CEA Saclay, IRAMIS, SPAM, Physique des Hautes Intensités, 91191 Gif-sur-Yvette Cedex, France - [3]Institut Néel, Dept MCMF, 38042 Grenoble, France.

E-mail: gerard.baldacchino@cea.fr

Abstract. Water radiolysis has been investigated by *in situ* and direct detection by using the scavenging method and 13-keV X-rays from ESRF synchrotron. By using a famous chemical system, concentrated hydrobromic acidic solutions over a range of concentrations (from 0.1 to 0.6 mole·dm^{-3}), and real time absorption spectroscopy of Br_3^- around 266 nm, we aimed at evaluating the effect of 13-keV X-rays, below the ionization K-edge of Br, on the Br^- oxidation yield value. The $HO^•$ scavenging time ranging in the picosecond scale is also taken into consideration going to the earliest initial yields. We have also observed the limit of use of N_2O and air as saturation gas. The oxidation limitation also comes from the presence of H_3O^+ in abundance giving birth of competitive reactions. The determination of the dose rate delivered to the solution has been performed by using the Fricke dosimeter system by following absorbance at 304 nm. The dose rate was 18.5 Gy·s^{-1}.

1. Introduction

Energetic X-rays are usually employed in many domains of science, industry and health care and investigation. Over a few 10 eV of energy, X-rays are ionizing particles with a Linear Energy Transfer (LET) similar to γ- rays (0.3 eV·nm^{-1}) [1]. The radiation chemistry in water generated by these rays are known and detailed in the literature [1,2]. In some cases such as dynamic structure of molecule under short time excitation [3], the effect of X-rays must not be neglected because the radicals produced have their own chemistry. In this way it becomes crucial to rationally study the effect of X-rays on the production of the most known oxidizing radical, namely the hydroxyl radical, $HO^•$. This species is largely involved in the oxidation of biological material like DNA, protein, membranes, etc and appears as the main species to investigate [4]. On the other hand, new system for the dosimetry is constantly required and a possibility recently investigated is the chemical system where hydrobromic acid (HBr) is involved at high concentration [5]. Bromine solutions have also been recently used to study the production of H2 in water radiolysis [6]. Nevertheless this chemical system has never been studied under X-rays. Moreover Br- is known for a long time as a good scavenger of hydroxyl radical[7] and its transient spectra and mechanism are known [7-11].

As an introduction to the radiolysis of HBr chemical system, we present in the table 1 the main reactions involved in the oxidation mechanism of Br^- by $HO^•$ in acidic conditions. In this article, the presented preliminary results concern a new study of HBr in high concentration conditions allowing the formation of the stable species Br_3^- and by using various gas saturations. The aim is first to determine the oxidation yield of Br^- by $HO^•$ with X-rays having an energy of 13 keV, below the Br

Journal of Physics: Conference Series **261** (2011) 012013 doi:10.1088/1742-6596/261/1/012013

ionization threshold [3,12]. In these conditions we have performed an optical setup as convenient as possible to detect the stable species formed: Br_3^-. It is then expected to obtain real-time and *in situ* spectra under X-rays irradiation. We also aim at introducing high-temperature experiments with HBr. Indeed, it is known that salts are precipating more efficiently at temperatures above 300°C. It is also obvious that the use of HBr solutions permit to avoid the cation effect (Na^+, K^+, Ga^{3+}) that can be seen in X-rays spectroscopies [13].

Table 1: Main reactions of the oxidation mechanism of bromide ions in acidic solutions

N°	Reactions			Rate constants[5]
1	$Br^- + HO^\bullet$	→	$BrOH^{-\bullet}$	$k_1 = 1.1 \times 10^{10}$ dm^3·mol^{-1}·s^{-1}
2	$BrOH^{-\bullet} + H_3O^+$	→	$Br^\bullet + 2H_2O$	$k_2 = 4.4 \times 10^{10}$ dm^3·mol^{-1}·s^{-1}
3	$Br^\bullet + Br^-$	→	$Br_2^{-\bullet}$	$k_3 = 1.2 \times 10^{10}$ dm^3·mol^{-1}·s^{-1}
4	$Br_2^{-\bullet} + Br_2^{-\bullet}$	→	$Br_3^- + Br^-$	$k_4 = 1.9 \times 10^9$ dm^3·mol^{-1}·s^{-1}
5	$Br_2 + Br^-$	⇆	Br_3^-	$K = 16.1$ dm^3·mol^{-1}

X-rays absorption spectroscopies at high temperature are also considered in a furtherer global study. That is the reason why a reference study of the oxidation yield of bromide ions under X-ray irradiation at room temperature is necessary.

2. Methods

2.1 Setup

The solutions were stable at room temperature under room light. All solutions were irradiated at room temperature (21 - 23°C) with X-rays from the synchrotron source of the European Synchrotron Radiation Facility (ESRF) at Grenoble (France). The ESRF synchrotron is an intense source with a large spectrum of photon emission. Inside the synchrotron ring, the beam mode chosen was 4 bunches. The time between 2 bunches is roughly 700 ns. The study presented here has been carried out with a synchrotron ring loaded with 4 bunches of electrons. The energy of the electron beamline is 6 GeV. A large range of photons energy is obtained when the 6 GeV-accelerated electrons go through a binding magnet in the beamline BM31 [14]. A monochromator selected an energy of 13 keV with a great accuracy. At this energy, the LET is 0.3 eV·nm^{-1}.The beam size is 200 × 300 μm².

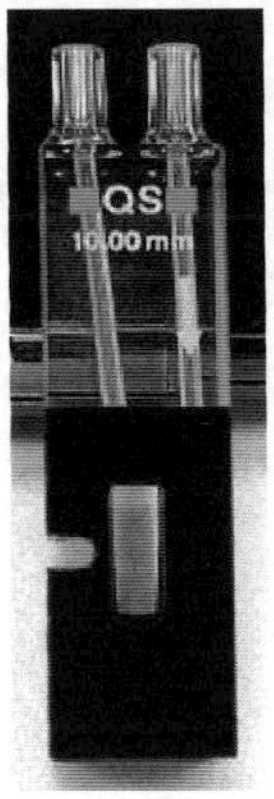

Figure 1. Commercial optical flow cell (Hellma ref. 176.000-QS) used in the irradiation setup with 13-keV X-rays. This cell is made of fused silica and the penetration of 13-keV photon was possible only by reducing the thickness of the window to 1 mm perpendicularly to the analysis optical axis. In the picture the machined window is located on the left side in the middle of the dark part. The size of the X-ray beam determined the optical path length to 300 μm.

A 13-keV X-ray beam crosses an optical flow cell (figure 1). Steady-state irradiations have been carried out by analyzing the temporal evolution of the absorption spectrum during short-time irradiations (a few tens of seconds at maximum).

The absorption spectra were recorded with a fiber-coupled CCD detector (Avantes, AVS-SD 2000 with Avasoft 5.1 software) operating without any synchronization but with the following parameters: 90 ms of integration time and 10 averaged scans. The spectra were periodically saved with 1 second between each spectrum. A reference spectrum before irradiation and a spectrum without light were performed in order to obtain directly the real time absorption spectrum. The light from a deuterium lamp (Avantes dual light source) is used to record the spectra in the UV domain (240 – 400 nm). This kind of setup has not really been developed by the radiation chemistry laboratories in the past (except some attempts such as the transient spectrum direct acquisition [15]). The advance in easy to use convenient CCD matrix, in sensitivities in the UV-range allows now to perform this detection without trouble and adjustment.

2.2 Chemicals

The aqueous solutions were prepared using ultra pure water from a Millipore system. The chemicals were used as received without further purification: HBr (purity of 62%) from Sigma-Aldrich. The pH of the HBr aqueous solutions was measured by using a Orion 9107WMMD electrode. The pH was acidic (pH = 0.4 - 1.0). The uncertainty on the measurement of the pH value was 0.1 unit of pH. The solutions were saturated with nitrous oxide (N_2O, purity > 99.98%) from Messer, by bubbling 30 minutes before irradiation.

2.3 Dosimetry

The dose rate used in the experiments was precisely determined by means of the Fricke method [16]. The chemical system is well described in the literature [1,16]. The Fricke dosimeter ($[Fe^{2+}] = 10^{-3}$ M, $[H_2SO_4] = 0.4$ M and $[NaCl] = 10^{-3}$ M) solution was prepared with $Fe(NH_4)_2(SO_4)_2(H_2O)_6$ from Merck, H_2SO_4 (purity > 99.999%) from Sigma-Aldrich and NaCl (purity > 99.8 %) from Labosi. Figure 2 shows the spectrum of the oxidized Fricke solution (Fe^{3+} spectrum) after 20 s of exposure.

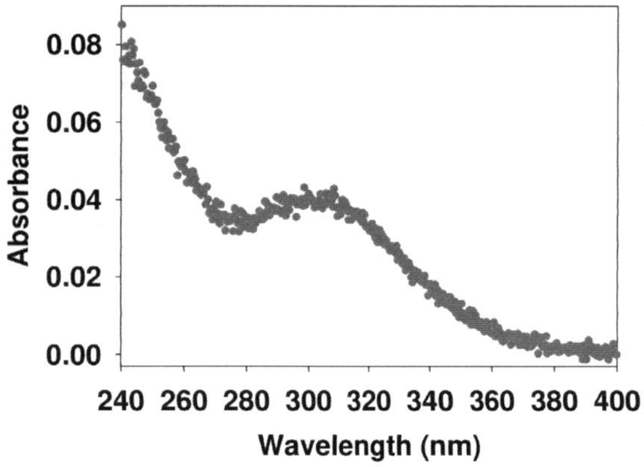

Figure 2. Absorption spectrum of the Fricke solution (1 mM of Fe^{2+}) after 20 s of 13-keV X-rays exposure. $\varepsilon_{304nm}(Fe^{3+}) = 2160\ M^{-1}\cdot cm^{-1}$. In this condition the dose at 20 s was 370 Gy falling in the known linearity domain of the Fricke dosimeter. At this stage dose rate was about 18.5 $Gy\cdot s^{-1}$.

The evolution of the absorbance at 304 nm is linear during at least 30 s as it can be noticed in the figure 3. The slope is determined with a good accuracy. Many Fricke dosimeter yields exist in the literature, especially for this range of X-ray energy [17,18]. The recommended value for 13-keV X-rays can be deduced from the article of J. Law *et al.*: 13.4 ions/100 eV (13.9×10^{-7} $mole\cdot J^{-1}$). The doses have been determined by using this value.

The challenge of knowing the dose rate with accuracy is linked to the knowledge of the intensity of the electrons which are in the storage ring. There is a tight correlation between the value of the dose rate and the intensity of the electrons which are accelerated in the ring.

Journal of Physics: Conference Series **261** (2011) 012013 doi:10.1088/1742-6596/261/1/012013

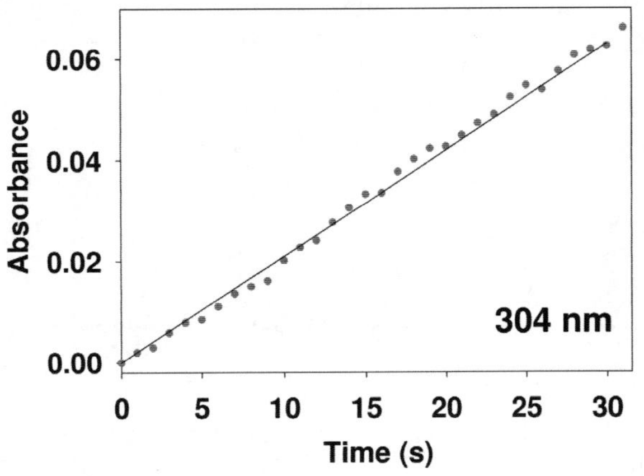

Figure 3. Absorbance evolution at 304 nm of a Fricke solution (1 mM of Fe^{2+}) during 13-keV X-rays irradiation. Optical pathlength of 300 μm. The slope of the linear adjustment corresponds to the dose rate. In these conditions the dose rate was 18.5 $Gy{\cdot}s^{-1}$.

The dose rate is calculated thanks to the slope of the linear domain of the Fricke dosimeter. This domain is limited to the primary phenomena. Thus, the dose rate is sensitive to the stability of the X-ray beam as soon as the exposure has begun. But during short-time irradiations (30 seconds), the monitored intensity of the X-ray beam gave relative variations which were inferior to the absorbance measurement accuracy. Therefore the dose rate can be determined and its value is 18.5 $Gy{\cdot}s^{-1}$.

3. Results and discussion

Several types of HBr aqueous solutions were irradiated with X-rays at similar dose rates. Solutions were aerated or saturated with N_2O gas. The concentrations of HBr were 0.1, 0.3 and 0.6 M. Results of a N_2O-saturated solution of HBr 0.1 M are presented in figure 4.

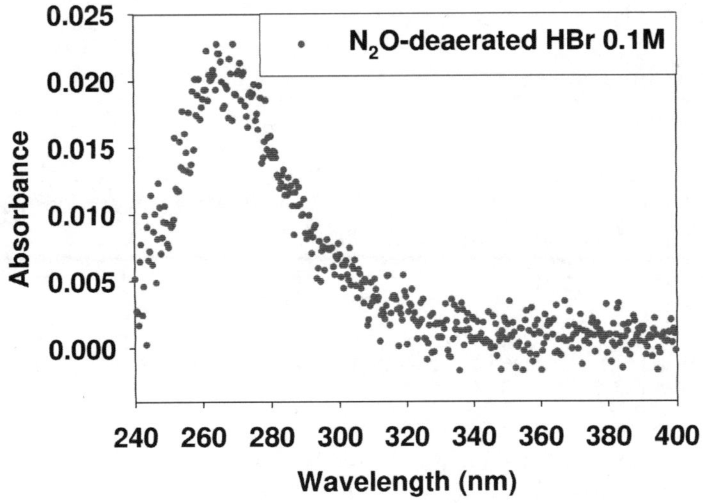

Figure 4. Absorption spectrum of a hydrobromic acid solution (0.1 M) after 10 s of 13-keV X-rays exposure. The solution was bubbled with N_2O and was not flushed into the cell during the irradiation.

Under X-rays exposure of a few seconds, an intense absorption band around 266 nm appeared. This absorption band corresponds to the formation of the stable tribromine ion, which is known to have a great molar extinction coefficient at this wavelength ($\varepsilon_{266nm}(Br_3^-)$ = 40900 ± 400 $M^{-1}{\cdot}cm^{-1}$). This intensity of the band at 266 nm is increasing with the irradiation time.

Considering the equilibrium between Br_2 and Br_3^- (reaction 5 in table 1), both species Br_2 and Br_3^- are present in solution. The proportions of each species with the initial concentration of bromide ions are gathered in table 2.

Table 2: Proportions of the oxidized forms of Br^- at three different pH values

HBr concentration (M)	pH	$[Br^-]$ (M)	$\%Br_3^-$	$\%Br_2$
0.1	1.0	0.1	61.7	38.3
0.3	0.7	0.2	76.3	23.7
0.6	0.4	0.4	86.6	13.4

Knowing that the molar extinction coefficient of Br_2 (47 $M^{-1}\cdot cm^{-1}$at 266 nm [19] is very low at this range of wavelength, we can conclude that the absorption which is visible at 266 nm, is mainly due to Br_3^-. The spectrum corresponds to the Br_3^- species which is stable enough and is accumulated in the cell. The optical pathlength was 300 μm.

The initial slope of the Br_3^- formation shown in figure 5 gives a value of the yield $G(Br_3^-)$ equal to $(0.9 \pm 0.1) \times 10^{-7}$ $mol\cdot J^{-1}$.

At this low pH (pH = 1.0), there is a tight competition between two fast reactions [20] which occur at different time scales:

$$e^-_{aq} + N_2O \rightarrow N_2 + O^{\bullet -} \qquad k_6 = 9.1 \times 10^9\ dm^3\cdot mol^{-1}\cdot s^{-1} \qquad (6)$$

$$e^-_{aq} + H_3O^+ \rightarrow H^\bullet + 2\,H_2O \qquad k_7 = 2.3 \times 10^{10}\ dm^3\cdot mol^{-1}\cdot s^{-1} \qquad (7)$$

The scavenging time of the reaction 6, if we consider the concentration of N_2O in our solution equal to $C_{N2O} = 2.5 \times 10^{-2}$ M at atmospheric pressure is $t_6 = 1/k_6C_{N2O} = 1/(9.1 \times 10^9 \times 2.5 \times 10^{-2}) = 4.4$ ns.

For a 0.1 M HBr solution, the scavenging time of the reaction 7 is $t_7 = 1/k_7C_{H3O+} = 1/(2.3 \times 10^{10} \times 0.1) = 0.4$ ns.

Considering the scavenging times of the reactions 6 and 7, roughly 10% of the produced hydrated electrons are converted into $HO^\bullet$. The largest part is converted into $H^\bullet$. Nevertheless, $H^\bullet$ produced by reaction 7 also contributes to the formation of $HO^\bullet$ according to reaction 8:

$$H^\bullet + N_2O \rightarrow N_2 + HO^\bullet \qquad k_8 = 7 \times 10^6\ dm^3\cdot mol^{-1}\cdot s^{-1}\ [20] \qquad (8)$$

This contribution plays an important role in the increase of the formation yield of the tribromine ion in the case of N_2O-saturated solutions of 0.1M HBr.

According to reaction 1 in table 1, in the presence of 0.1 M HBr, the $HO^\bullet$ scavenging time is 0.9 ns. Thus, the $HO^\bullet$ scavenging by bromide ions is favored over the recombination reaction:

$$HO^\bullet + HO^\bullet \rightarrow H_2O_2 \qquad k_9 = 4.2 \times 10^9\ dm^3\cdot mol^{-1}\cdot s^{-1}\ [20] \qquad (9)$$

Both reactions 2 and 3 happening quickly, $BrOH^{\bullet -}$ mainly reacts first with H_3O^+, which is abundant in the solution, to lead to the species $Br^\bullet$. This species reacts quickly with a bromide ion to give $Br_2^{\bullet -}$. The disproportionation reaction of $Br_2^{\bullet -}$ leads to the formation of the tribromine ion (reaction 4). However, this last reaction is in competition with another fast one which corresponds to the destruction of the precursor species $Br_2^{\bullet-}$ by hydrogen atoms (reaction 10) produced by the water radiolysis and mainly thanks to the reaction 7:

$$Br_2^{\bullet-} + H^{\bullet} \rightarrow 2\,Br^- + H^+ \qquad k_{10} = 1.4 \times 10^{10}\,dm^3{\cdot}mol^{-1}{\cdot}s^{-1} \qquad (10)$$

The formed tribromine ion is also consumed by $H^{\bullet}$ following a fast reaction :

$$Br_3^- + H^{\bullet} \rightarrow Br_2^{\bullet-} + Br^- + H^+ \qquad k_{11} = 1.2 \times 10^{10}\,dm^3{\cdot}mol^{-1}{\cdot}s^{-1} \qquad (11)$$

These reactions 10 and 11 are a reason to explain why the formation yield of Br_3^- is slightly reduced after an exposure time superior to 10 seconds.

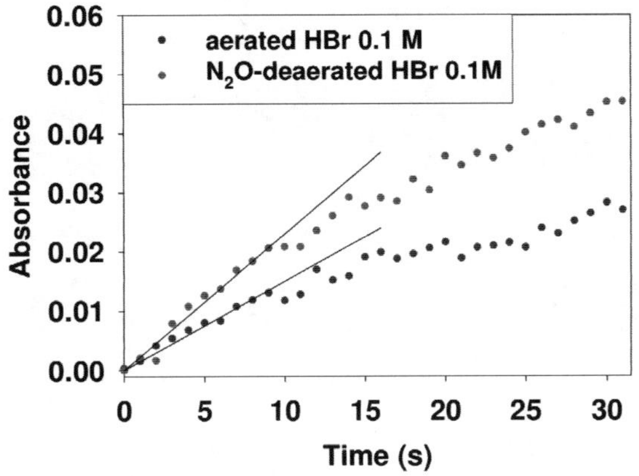

Figure 5. Absorbance at 266 nm of a hydrobromic acid solution (0.1 M) under 13-keV X-rays. The solution was aerated or bubbled with N_2O and was not flushed into the cell during the irradiation.

On the figure 5, we observe that there is an increase of the absorbance at 266 nm with the exposure time for both considered gas saturations. We notice that the variation is no more linear, the formation yield of Br_3^- decreasing with the exposure time.

Under N_2O, the increase of the absorbance due to Br_3^- is slightly more important than for the aerated solution. It seems that the impact of $H^{\bullet}$ on the destruction of $Br_2^{\bullet-}$ and Br_3^- is limited in the case of N_2O. As previously seen, for N_2O-saturated solutions, a small part of the hydrated electrons are converted into $HO^{\bullet}$ (reaction 8). These hydroxyl radicals participate to the oxidation mechanism by scavenging bromide ions and then contribute to increase the formation yield of Br_3^-.

In an aerated medium and at this pH value (pH = 1.0), the formation yield of Br_3^- is found to be lower than the yield determined for the N_2O-saturated solution. The hydrated electrons are totally converted into $H^{\bullet}$. The generated hydrogen atoms react with O_2 present in the solution (2.5×10^{-4} M at atmospheric pressure):

$$H^{\bullet} + O_2 \rightarrow HO_2^{\bullet} \qquad k_{12} = 1.3 \times 10^{10}\,dm^3{\cdot}mol^{-1}{\cdot}s^{-1} \qquad (12)$$

This production of $HO_2^{\bullet}$ leads to a decrease of the Br_3^- formation yield because $HO_2^{\bullet}$ reduces all Br^- oxidized forms. The rate constant of these reactions is quite important ($1.0 \times 10^8\ dm^3{\cdot}mol^{-1}{\cdot}s^{-1}$).

Table 3 gathers values of the Br_3^- formation yield for different HBr concentrations of aerated and N_2O-saturated solutions. In the first column, the yield values have been obtained thanks to the slopes of the figure 5. In order to observe the influence of the initial concentration of bromide ions, we have worked at higher concentrations: 0.2 M and 0.4 M. A few experimental and simulation values of the $HO^{\bullet}$ yield can be found in the literature [3,21].

In this study, a relation between $G(Br_3^-)$ and $G(HO^\bullet)$ cannot be anticipated because the scavenging time is short and several reactions of formation and destruction can occur in a complex mechanism. The best way could be simulation calculations including short-time and long-time scale.

Table 3: Effect of the HBr initial concentration on the Br_3^- formation yield under 13-keV X-rays

HBr concentration (mole·dm^{-3})	0.1	0.3	0.6
Scavenging time of Br^- (ns)	0.9	0.5	0.2
$G(Br_3^-)$ for aerated solutions (10^{-7} mol·J^{-1})	0.6 ± 0.1	1.6 ± 0.2	2.6 ± 0.3
$G(Br_3^-)$ for N_2O-saturated solutions (10^{-7} mol·J^{-1})	0.9 ± 0.1	1.0 ± 0.2	2.1 ± 0.3
Absorbance due to Br_3^- after 30 seconds of exposure for aerated solutions	0.03	0.08	0.105
Absorbance due to Br_3^- after 30 seconds of exposure for N_2O-saturated solutions	0.045	0.055	0.075

First, in the case of Br^- aerated solutions, the formation yield of the tribromine ion increases constantly with the Br^- concentration to reach $(2.6 \pm 0.3) \times 10^{-7}$ mol·J^{-1} which corresponds to the expected effect widely described in the literature [3,19]. The formation of Br_3^- is also favored at high concentrations in bromide ions if we consider the equilibrium between Br_3^- and Br_2 (reaction 5). An increase of the initial Br^- concentration leads to an increase of the ratio between Br_3^- and Br_2 (table 2). Thus, the absorbance at 266 nm due to Br_3^- is higher. In presence of N_2O, the effect is similar and the Br_3^- formation yield reaches $(2.1 \pm 0.3) \times 10^{-7}$ mol·J^{-1}. Whereas $G(Br_3^-)$ for N_2O-saturated 0.1 M HBr solutions is higher than $G(Br_3^-)$ for aerated 0.1 M HBr solutions (first column in table 3), we notice that in case of 0.6 M HBr, the Br_3^- formation yield found under N_2O is slightly inferior to the value found for the aerated system (third column in table 3).

The concentration of H_3O^+ increases with the concentration of bromide ions, so the hydrated electrons are scavenged much rapidly by H_3O^+ to lead to H-atoms. For example, the scavenging time of e^-_{aq} by H_3O^+ (reaction 7) is 110 ps for 0.6 M HBr whereas it is 430 ps for 0.1 M HBr. It means that the proportion of hydrated electrons which are converted into $HO^\bullet$ radicals by N_2O (reaction 6) is only 3%. In other terms, the effect of N_2O in the medium becomes negligible.

In order to describe the pH effect on the oxidation mechanism of bromide ions and to understand why the Br_3^- formation is not linear with the 13-keV X-rays exposure time, it is crucial to highlight the main reactions involved in both gas saturation conditions: N_2O-saturated and aerated aqueous solutions. In the case of a N_2O-saturated 0.6 M HBr solution ($[Br^-] = 0.4$ M, pH = 0.4), we have to consider the reactions in which $H^\bullet$ are involved because these radicals are present in a large amount in our acidic conditions. As previously seen, $H^\bullet$ can react with N_2O (reaction 8), with Br_3^- (reaction 11) and also with Br^- (reaction 13):

$$H^\bullet + Br^- \rightarrow HBr^{\bullet-} \qquad k_{13} = 1.7 \times 10^6\, dm^3{\cdot}mol^{-1}{\cdot}s^{-1} \qquad (13)$$

At 0.4 M of Br^-, 97% of the hydrated electrons are converted into $H^•$. The scavenging time of the reaction 13 is the shortest one (t_{13} = 1.8 μs). The reaction 11, which consumes Br_3^-, is in competition with reaction 13 as soon as the Br_3^- concentration reaches a concentration of 5×10^{-5} mol·dm^{-3}. It corresponds to an absorbance of 0.06. According to table 3, the absorbance of the N_2O-saturated 0.6 M HBr solution at the end of the exposure was 0.075. This value reveals that the reactions 11 and 13 are in a tight competition after about 10 seconds of irradiation.

In the case of an aerated 0.6 M HBr solution, the reactions 11 and 13 are also highlighted as in the case of a N_2O-saturated solution. In addition, we have to take into account the reaction between $H^•$ and O_2, which is present in the atmosphere (reaction 12). The scavenging time of this reaction 12 is the shortest one (t_{12} = 0.3 μs). The competition between the reactions 12 and 13 happens then when the concentration of the formed tribromine ion reaches 3.0×10^{-4} mol·dm^{-3}, which corresponds to an absorbance of 0.3. According to table 3, the absorbance of the aerated 0.6 M HBr solution at the end of the irradiation was 0.105. This low value shows that the competition does not play an important role within the 30 seconds of exposure. The competition should be more significant after 60 seconds of irradiation. However, the decrease of the Br_3^- formation yield is visible. It could be partially due to $HO_2^•$. Indeed, this radical is produced by the reaction 12 and reacts quickly with $Br_2^{•-}$ and Br_3^- with similar rate constants, 1.0×10^8 dm^3·mol^{-1}·s^{-1}.

In both cases, we observe that the diminution of the oxidation yield is limited. Despite the destruction of Br_3^- by $H^•$, we notice, on the one hand, there is a non negligible production of oxidizing species ($HO^•$, $HBr^{•-}$) which contributes to raise this yield in the case of N_2O-saturated HBr solutions. On the other hand, there is the presence of the molecular oxygen which scavenges a part of $H^•$ and limits their impact in the case of aerated HBr solutions.

4. Conclusion

Those preliminary results obtained on the ESRF synchrotron are a great opportunity to understand better the oxidation mechanism of bromide ions under X-rays irradiation. The analysis of the *in situ* formation of the tribromine ion could give information about the radiolytic yield of $HO^•$.

Unfortunately, $HO^•$ radical yield is not easily accessible from the Br_3^- formation resulting from the scavenging of $HO^•$ by Br^-. As the complexity of the chemical system does not allow us to determine the $HO^•$ yield, it should be interesting to carry out simulation in a large time scale (from ps to μs) in order to take into account the scavenging processes in the spurs and the homogeneous chemistry for the reactions of secondary products ($Br_2^{•-}$, $Br^•$...). Actually, the use of HBr system, initially motivated by the absence of the counter-ion unlike in the use of salts NaBr or KBr, implies a strong effect of the pH. This paper was thus an opportunity to detail the chemical mechanism of Br^- action in presence of H_3O^+.

The concentrations of hydrobromic acid were 0.1, 0.3 and 0.6 M, corresponding to three scavenging times. The effect of two saturation gas was also used: aerated and N_2O saturation. The presence of N_2O in the aqueous bromide solutions has shown an increase in the value radiolytic yield of Br_3^- comparing to the yield determined for the aerated solution but only for the lowest concentration of HBr, 0.1 M. This increase is due to the part of the hydrated electrons which is converted into $HO^•$ and thus, which contributes to the oxidation of the bromide ions.

At higher concentrations of HBr (0.3 and 0.6 M), as expected, the oxidation yield of Br^- aims at increasing. It was explained by two main reasons. First, by increasing the concentration of Br^-, the $HO^•$ scavenging power is enhanced. So, the oxidation yield of Br^- is raised following the higher and higher $HO^•$ yield. Then, considering the equilibrium which is present in the oxidation mechanism, by raising the initial concentration of Br^-, the ratio between Br_3^- and Br_2 is greater and the absorbance at 266 nm due to Br_3^- is higher. Nevertheless, the presence of N_2O, unlike O_2 in aerated solutions, complicates the system. N_2O should normally aims at increasing the tribromine ion yield but the competition between reactions existing in the case of the presence of O_2, does not exist at all in N_2O saturation. H-atoms which are involved in the mechanism are then available to react with $Br_2^{•-}$ and influence the decrease way of the observed yield of Br_3^-. Therefore the initial pH value plays an important role in the understanding of the mechanism too. At acidic pH, hydrated electrons are rapidly converted into $H^{\bullet}$. This study has then exhausted a system of several reactions in competition involving $H^{\bullet}$ which is less pronounced in the case of neutral NaBr and KBr solutions.

This work has been carried out with a 13-keV X-rays beam which corresponds to energy below the Br atom ionization threshold. It could be interesting to work at higher X-rays energies in order to observe the effect of the direct ionization of Br atom on the oxidizing mechanism.

Finally this study stresses the importance of the liquid state radiolysis under X-rays. Several spectroscopic techniques using X-rays do not take into account the effect of this ionizing ray on the degradation of the samples. It clearly shows that the radiolysis is a phenomenon which cannot be neglected.

References

[1] Spinks J W T and Wood R J *An Introduction to Radiation Chemistry* Wiley 1990.
[2] Mozumder A and Hatano Y *Charged Particle and Photon Interactions with Matter. Chemical, Physicochemical, and Biological Consequences with applications*, Marcel Dekker, 2004.
[3] Elles C G, Shkrob I A , Crowell R A , Arms D A and Landahl E C 2008 *J. Chem. Phys.* **128** 061102.
[4] J. Belloni, M. Mostafavi, T. Douki, M. Spotheim-Maurizot 2008 L'*Actualité Chimique* **316** I.
[5] Mirdamadi-Esfahani M , Lampre I , Marignier J L , de Waele V and Mostafavi M 2009 *Rad. Phys. Chem.* 78 106.
[6] LaVerne J A, Ryan M R and Mu T T 2009 *Rad. Phys. Chem.***78** 1148.
[7] Zehavi D and Rabani J 1972 *J.Phys. Chem.***76** 312.
[8] Hug G L *Optical Spectra of Nonmetallic Inorganic Transient Speciec in Aqueous Solution*, NBS, 1981.
[9] Behar D 1972 *J.Phys.Chem.* **76** 1815.
[10] Mamou A, Rabani J and Behar D 1977 *J. Phys. Chem.* **81** 1447.
[11] Matheson M S, Mulac W A, Weeks J L and Rabani J 1966 *J. Phys. Chem.* **70** 2092.
[12] Uno K, Notoya Y, Fujikawa T, Yoshikawa H and Nishikawa K 2005 *Japan. J. Appl Phys. Part 1* **44** 4073.
[13] Da Silva C, Proux O, Hazemann J L, James-Smith J, Testemale D and Yamaguchi T 2009 *J. Mol. Liq.* **147** 83.
[14] O. Proux O, Biquard X, Lahera E, Menthonnex J J, Prat A, Ulrich O, Soldo Y, Trevisson P, Kapoujyan G, Perroux G, Taunier P, Grand D, Jeantet P, Deleglise M, Roux J P and Hazemann J L 2005 *Phys. Scripta* **T115** 970.
[15] Baldacchino G and Hickel B 1998 *Rev. Sci. Inst.***69** 1605.
[16] Fricke H and Hart E J *Radiation chemistry*, Vol. II, Academic Press, 1966.
[17] Frilley M and Lefort M 1959 *J.Phys.Le Radium* **20** 516.
[18] Law J 1969 *Phys. Med. Biol..* **14** 607.
[19] Beckwith R C, Wang T X and Margerum D W 1966 *Inorg. Chem.* **35** 995.
[20] Buxton G V, Greenstock C L, Helman W P and Ross A B 1988 *J. Phys. Chem. Ref. Data* **17** 513.
[21] Jonah C D and Miller J R 1977 *J.Phys. Chem.* **81** 1974.

AUTHOR INDEX

Atinault, E 132
Baldacchino, G 132
Banyasz, A 85
Barré, L 108
Bauer, G 1
Bergès, J 25
Beuve, M 67
Bournel, F 78
Chapon, F 108
Collet, S 108
Colliaux, A 67
Confalonieri, F 40
Constans, J M 108
Courtheoux, P 108
Davídková, M 98
Derlon, J M 108
Dou, W 108
Drouin, R 15
Fong, V 108
Francesconi, S 15
Gauduel, Y A 55, 108
Gervais, B 67
Girard, P M 15
Graindorge, D 15
Guillamo, J S 108
Gustavsson, T 85
Hazemann, J L 132
Hossu, G 108
Houée-Levin, C 25
Improta, R 85
Kauffmann, F 108
Khodja, H 125
Kobayashi, K 33
Lacombe, S 33, 78, 108
Laffon, C 78
Lasne, J 78
Le Sech, C 33
Lechapt-Zalcman, E 108
Levin, C Houée 108
Markovitsi, D 85
Parent, Ph 78
Pin, S 132
Porcel, E 33
Pozzebon, M 15
Remita, H 33
Renault, J P 132
Rioult, F 108
Rochette, P 15
Rodriguez-Lafrasse, C 67
Ruan, S 108
Saffré, D 132
Sage, E 15
Sommer, S 40
Spotheim-Maurizot, M 98
Trouillas, P 25
Usami, N 33
Valable, S 108